SpringerBriefs in Applied Sciences and Technology

SpringerBriefs present concise summaries of cutting-edge research and practical applications across a wide spectrum of fields. Featuring compact volumes of 50 to 125 pages, the series covers a range of content from professional to academic. Typical publications can be:

- A timely report of state-of-the art methods
- An introduction to or a manual for the application of mathematical or computer techniques
- A bridge between new research results, as published in journal articles
- A snapshot of a hot or emerging topic
- An in-depth case study
- A presentation of core concepts that students must understand in order to make independent contributions

SpringerBriefs are characterized by fast, global electronic dissemination, standard publishing contracts, standardized manuscript preparation and formatting guidelines, and expedited production schedules.

On the one hand, **SpringerBriefs in Applied Sciences and Technology** are devoted to the publication of fundamentals and applications within the different classical engineering disciplines as well as in interdisciplinary fields that recently emerged between these areas. On the other hand, as the boundary separating fundamental research and applied technology is more and more dissolving, this series is particularly open to trans-disciplinary topics between fundamental science and engineering.

Indexed by EI-Compendex, SCOPUS and Springerlink.

More information about this series at http://www.springer.com/series/8884

Rucha Joshi

Collagen Biografts for Tunable Drug Delivery

 Springer

Rucha Joshi (iD)
Biomedical Engineering
University of California, Davis
Davis, CA, USA

ISSN 2191-530X   ISSN 2191-5318   (electronic)
SpringerBriefs in Applied Sciences and Technology
ISBN 978-3-030-63816-0   ISBN 978-3-030-63817-7   (eBook)
https://doi.org/10.1007/978-3-030-63817-7

This Springer imprint is published by the registered company Springer Nature Switzerland AG
The registered company address is: Gewerbestrasse 11, 6330 Cham, Switzerland

I dedicate this book to my energy-infused Ph.D. advisor Dr. Sherry L. Voytik-Harbin, who inspired me to go on an exciting journey of exploring the world of collagen; to my beloved husband Dr. Shashank Tamaskar, who has been a constant reservoir of calmness and charm during the completion of this project; and finally to my parents – Deepa and Vinay, and my daughter Anvi, who have constantly been my pillars of strength.

Preface

One of the biggest challenges in tissue engineering currently is the formation of a functional microvascular network as part of an engineered tissue graft. Despite many advances in tissue engineering methods, the field still awaits biograft designs that enable neovascularization at clinically relevant size scales. Critical to the design of such materials are tissue-specific physico-mechanical properties and controlled local therapeutic molecular release.

The purpose of this book is to review collagen-based biomaterials that have been applied broadly to tissue engineering and local drug delivery applications and lay-out a landscape for developing a multifunctional biograft material from collagen polymers. Type I collagen forms an ideal candidate for tissue engineering and drug delivery application because of the several advantages it possesses over other materials, such as its high availability in the body, conservation across tissue and species, inherent self-assembly, biological signaling capacities, proteolytic biodegradability, and low immunogenicity. Type I collagen fibril-matrix can be formulated as an implantable or injectable drug delivery device and is known to provide scaffolded support to the cells, while organizing the cells three-dimensionally, and impart essential information to cells to proliferate/migrate and to regulate their behavior.

This book discusses the current collagen-based drug delivery landscape as well as new opportunities for forming a multifunctional collagen-based drug delivery device for soft tissue construction. Chapter 1 establishes a background of collagen-based drug delivery devices. Chapter 2 discusses how a successful drug delivery device can be developed from collagen by using a two-fold approach: (i) choosing an appropriate method of isolating collagen, such that it does not damage the native structure of the collagen molecule and its properties; and (ii) choosing an appropriate drug incorporation strategy, such that maximum drug can be loaded without chemically altering the native structure of collagen. Chapter 3 identifies a few novel strategies for tuning molecular release from collagen. Chapter 4 reviews the application of collagen biograft materials to chronic wounds. Finally, in Chap. 5, the use of heparin for affinity-based vascular endothelial growth factor (VEGF) retention in collagen constructs is discussed for promoting neovascularization. An established *in-vivo* chicken egg chorioallantoic membrane (CAM) model is also presented for evaluating a collagen-based drug delivery device's ability to promote neovascularization and tissue regeneration.

It is my sincere hope that this book will illuminate the reader on the current collagen-based drug delivery landscape, advantages, and limitations of conventional collagen-based products, and inspire a formation of porous collagen materials resembling those found in tissues, which can maintain their inherent biological signaling properties while providing an ideal multifunctional platform for integrated tissue engineering and molecular therapy design.

Davis, CA, USA   Rucha Joshi

Contents

About the Author

Rucha Joshi, Ph.D. is an assistant professor in the Department of Biomedical Engineering at the University of California, Davis, focusing on engineering education research and instructional innovation in biomedical engineering. Prior to joining UC Davis in 2018, she was a postdoctoral fellow in the Weldon School of Biomedical Engineering, Purdue University, working on multiple educational projects in enhancing teaching, learning, outreach, and diversity of engineers. Rucha's current research focuses on approaching challenges in teaching engineering through the lens of design thinking. Previously, Rucha contributed to instructional innovation in biomedical engineering at Purdue and worked on an NSF-funded grant for studying the professional formation of engineers and enhancing diversity and inclusion within Purdue. Rucha is also actively involved in educational entrepreneurship projects and making engineering accessible to underrepresented high school students in the United States, as well as India.

Rucha received a BS in biotechnology engineering from Shivaji University, India, in 2009, and MS in biomedical engineering from Vanderbilt University in 2011, where she worked on creating smart microspheres for drug delivery with Dr. Craig Duvall. At Vanderbilt, she designed and developed a dual temperature and pH-responsive microsphere delivery system for sustained protein delivery to ischemic environments. She also synthesized and characterized microparticle-based delivery system platform from reactive oxygen species (ROS) responsive poly(propylene sulfide) (PPS) for sustained drug delivery. Rucha received her PhD in biomedical engineering from Purdue University in 2016, where she worked on creating designer collagen biografts for enabling controlled drug delivery. Rucha developed multifunctional 3D collagen-fibril biograft materials with tunable physical and molecular delivery properties with Prof. Sherry Voytik-Harbin. She also devised and validated an in vitro model system for quantifying kinetics of molecular release from collagen, and applied collagen biograft materials for enhanced local neovascularization *in vitro* and *in vivo* CAM model.

Rucha has received several scholarships and awards for thought leadership, entrepreneurship, and community engagement, including the Ross Fellowship for outstanding PhD applicants, and the Burton D. Morgan Fellowship for entrepreneurial students at Purdue. She was honored by Dr. APJ Abdul Kalam, the late president of India, for her invention of "low calorie biscuits from banana peel pulp" and for her STEM popularization efforts. Rucha authored a book in Marathi, *Bhartiya Balvaidnyanik achi Garudzep*, summarizing her life experiences in inspiring Indian students towards STEM pursuits, which won a literary award from the State Government of Maharashtra.

Chapter 1
Introduction to Drug Delivery

1.1 Introduction

Rapid growth in the North American drug delivery market has been driven by increased adoption of drug delivery technologies by pharmaceutical and medical device companies, and greater patient acceptability of drug delivery products. It has been recognized that the improvements in drug therapy are a consequence of not only a new chemical entity but of a combination of active substance and delivery system. However, developing a new chemical entity is more expensive and time-consuming than a novel drug delivery technology. The average cost and time required to develop a new drug is approximately $500 million and over 10 years, which is significantly higher than that for developing a new drug delivery system (approximately $20–50 million and 3–4 years) [1]. Clearly, the added value of the latter has made several pharmaceutical companies employ drug delivery technology as a life cycle management tool for some of their blockbuster drugs.

Drug delivery technologies enable modulation of the time course of drug release and make localization of therapeutic agents possible. The main function of drug delivery system is to transport a drug or biologically active agents such as growth factors and genetic material [2] into a desired location in order to promote the therapeutic treatment of disorders and diseases [3]. The collagen-based therapeutic delivery device may include one active agent or, optionally, more than one active agent. The one or more active agent included in the device also can vary depending on the intended purpose, location, and placement of the device. As used herein, the term "active agent" refers to any compound, agent, molecule, biomolecule, drug, therapeutic agent, nanoparticle, peptide, protein, polypeptide, antibody, ligand, partial antibody, steroid, growth factor, transcription factor, DNA, RNA, virus, bacteria, lipid, vitamin, small molecule, or large molecule that has activity in a biological system. Active agents include but are not limited to "biomolecules," "drugs," silver amine complexes, surfactants, polyhexamethylenebiguanidine, betaine, antimicrobials, linear polymer biguanidines with a germicidal activity, bioactive additives, heparin, glyosaminoglycans, extracellular matrix proteins, antibiotics, growth factors, epidermal growth factor (EGF), platelet-derived growth factor (PDGF), fibroblast growth factor (FGF), collagen-binding peptides or factors, connective tissue activating peptides (CTAP), transforming growth factors (TGFs), oncostatic agents, immunomodulators, immunomodulating agents, anti-inflammatory agents, osteogenic agents, hematopoietic agents, hematopoietic modulators, osteoinductive agents, TGF-β1, TGF-β2, TGFα, insulin-like growth factors (IGFs), tumor necrosis factors (TNFs), interleukins, IL-1, IL-2, colony-stimulating factors (CSF), G-CSF, GM-CSF, erythropoietin, nerve growth factors (NGF), interferons (IFN), IFN-α, IFN-β, IFN-γ, preservatives, dyes, non-bioactive agents, hormones, as well as synthetic analogs of the above.

After the past four decades of research and remarkable progress in the development of drug delivery systems [4], the focus of designing drug delivery system has shifted from "sustained release" to "controlled release." "Sustained release" technologies dealt with simple prolongation of drug in circulating blood within the appropriate therapeutic range as the satisfactory endpoint [5]. However, the need for precise control and minimization of intra- and inter-subject variability normally associated with drug delivery systems has led to the development of "controlled release" technologies.

Controlled drug delivery is a technology that presents a drug to the target site in the body in a spatiotemporally controlled manner, while simultaneously attempting to protect the drug from physiological degradation or elimination. Saltzman [6] declared that a controlled release drug delivery system must (1) include a component that can be engineered to regulate essential characteristics (such as drug release rate or targeting) and (2) have a duration of action longer than a day. Today, controlled drug delivery in which drug efficacy can be influenced and controlled by the delivery vehicle through manipulation of drug concentration, as well as its spatiotemporal properties (i.e., localization of drug, length of circulation time), has developed to the point where it is applicable to almost every field in medicine and every organ system, including the heart [7], brain [8, 9], eye [10], and peritoneum [11]. Devices were developed that could release compounds with high reproducibility [12] and constant release rates [13].

While drug delivery is thus a promising application of biomaterials, in many of the medical cases when patients suffer with trauma, burns, tumor resection, congenital defects, and chronic wounds a biomaterial is needed to do more than just deliver a drug molecule. The biomaterial needs to offer multi-scale structure and function to support soft tissue reconstruction in those cases. Current gold-standard technology includes options such as using autografts; however, due to the lack or limited supply of autograft tissues, treating patients in need of soft tissue reconstruction still remains a major surgical challenge [14]. This supply-demand gap (Fig. 1.1) is yet to be filled up. Various types of scaffolds prepared from synthetic or natural materials have been used for simple reconstruction, but their effectiveness remains limited by slow neovascularization ultimately contributing to poor functional integration, pain, and/or scarring. An ideal solution for clinicians would be a designer biologic graft material that provides appropriate multi-scale structure and function while fostering rapid vascularization for improved tissue integration and regeneration.

Fig. 1.1 Cartoon depicting the gap between autograft supply and demand for soft tissue reconstruction

In considering biomaterials that can be used for integrated tissue engineering and local molecular delivery strategies targeting graft vascularization, some key design criteria need to be satisfied [15]. An ideal platform biomaterial should be mechanically strong and stable, allowing for cell co-implantation and cell infiltration. Furthermore, it should be able to support structurally and functionally support cell growth. Finally, it should be able to fully integrate with the tissue physiologically, thus degrading at a rate that supports new tissue regeneration. From a drug delivery perspective, it must also be able to support a broad range of customizable spatiotemporal molecular release profiles. This could be especially advantageous for growth factor delivery to support neovascularization and accelerated tissue regeneration at the site of application. Finally, an ideal biomaterial should not impede the growth of regenerating tissue and its degradation should not leave behind any byproducts that adversely affect the cells involved in tissue regeneration. To summarize, an ideal drug delivery platform for soft tissue reconstruction would therefore entail the development of a multifunctional soft tissue graft material that provides (1) tissue-specific physico-mechanical properties and (2) controlled, local therapeutic molecular release for accelerated neovascularization and functional tissue integration and regeneration (Fig. 1.2). We focus on this merger of drug delivery technology with biomaterial platforms in this book.

1.2 Wonder-Material Collagen

Among various natural and synthetic biomaterials that can be considered candidate materials for soft tissue reconstruction, type I collagen provides an attractive choice, as it is the major structural and mechanical component of the majority of connective tissues and organs. Until now, 29 different collagen types have been identified and classified into distinct groups [16] based on their structure and localization within specific tissues. However, [17], type I collagen is the most well-studied and most commonly used in drug delivery because it is the major constituent present in virtually every tissue.

Collagen accounts for more than 90% of the extracellular matrix (ECM) in skin, bone, and tendon of vertebrates [18] and approximately 30% of total body protein [19, 20]. Its ability to form polymerizable, porous collagen-fibril matrices in vitro that can degrade into physiologically well-tolerated products make it an excellent biocompatible material with low immunogenicity. Additionally, its versatility and ability to be processed on an aqueous basis make it a viable candidate for formulat-

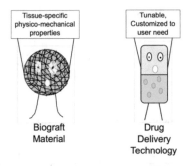

Fig. 1.2 Schematics showing that a functional tissue regeneration requires merger of a multifunctional biograft material with tunable drug delivery technology that can affect solute/fluid transport and ultimately the cell fate and tissue formation

ing drug delivery systems [16]. Being a natural polymer, collagen also provides advantages related to its inherent cell-signaling potential which is facilitated by adhesion domains that engage in integrin-mediated cell binding. In fact, type I collagen stands out as one of the most pro-angiogenic super-polymers due to its suitability as a substrate for endothelial cell attachment and signaling. Biophysical cues defined by the hierarchical assembly and intermolecular cross-linking of collagen molecules add to its attractiveness for drug delivery applications. Type I collagen, the predominant and major structural component of the Extra Cellular Matrix (ECM) thus represents an ideal natural polymer candidate for an integrated tissue engineering and local molecular delivery strategies [19].

One would think that with all these ideal properties, collagen should be widely applied clinically and largely available commercially. However, to our dismay, despite the numerous advantages of type I collagen as a natural biomaterial, its use as a vehicle for controlling molecular release is found to be limited. Furthermore, its application as a tissue graft that induces tissue regeneration while simultaneously providing predicted, localized delivery of drug molecules has not been achieved successfully on a large, commercial scale. In fact, a very few collagen-based active agent delivery formulations have made it to commercial availability as indicated in Table 1.1. What is the reason behind this under-utilization of type I collagen in clinics and in market?

The marginal success of these present-day collagen-based drug delivery formulations can be traced to their major shortcoming in that they are poorly characterized in terms of their molecular composition and none fully capitalize on the inherent

Table 1.1 Examples of collagen-based drug delivery products in market

Drug delivery system	Company	Drug incorporated	Dressing format	Application
Cogenzia	Innocoll	Gentamicin	Lyophilized collagen sponge	Treatment of diabetic foot infections
XaraColl	Innocoll	Bupivacaine hydrochloride		Local anesthetic sponge for postoperative pain relief
Vitagel™	Orthovita Inc.	Thrombin	Suspension of bovine collagen and bovine thrombin in $CaCl_2$ buffer	Surgical hemostatic gel
ColActive® Plus Ag	Covalon	EDTA and silver ions	Lyophilized collagen sponge made with collagen, carboxyl methylcellulose (CMC), and sodium alginate	Chronic wound healing
Promogran Prisma™ Ag	Acelity	Silver-ORC containing 25% w/w ionically bound silver (Ag)	Lyophilized sponge consisting of 44% oxidized regenerated cellulose (ORC), 55% collagen, and 1% silver-ORC	Healing chronic wounds such as diabetic venous and pressure ulcers
Biostep Ag	Smith and Nephew	EDTA and silver (Ag)	Lyophilized sponge made from porcine type I collagen and gelatin	Healing chronic wounds such as diabetic, venous, and pressure ulcers, and burns

Source: Company based literature and BCC Market Research [21]

self-assembly or polymerization capacity of collagen. The existing collagen formulations are generally categorized into either non-dissociated fibrillar collagens or solubilized collagens (which we expand upon in Chap. 2). These current formulations have several shortcomings which are discussed briefly below.

The conventional formulations exhibit amorphous microstructures, with cursory control of material properties, including pore size and proteolytic degradability, through modulation of lyophilization conditions and/or exogenous chemical and physical crosslinking. Materials formed without any cross-linking are characterized as mechanically unstable, too soft to handle, and unable to resist cell-induced contractions [20, 22–24] thus failing to support cell ingrowth and migration required for tissue regeneration. On the other hand, exogenous crosslinking [20, 25–31] has been shown to have detrimental effects on cells and tissues [32], such as cytotoxicity [33, 34] or tissue calcification [35–37] and partial denaturation of collagen itself [26, 38]. When such formulations are used for drug delivery, problems related to mechanical integrity, and inability to give controlled release without cross-linking, are observed. They limit the clinical success of collagen for tissue engineering and molecular delivery applications [39–41]. Thus, conventional collagen formulations' inadequacies including poorly defined molecular composition, low mechanical integrity, rapid biodegradation, and limited control over drug release profiles have endangered the application of collagen in clinical use.

For tapping into the true potential of collagen for multifunctional, soft tissue reconstruction and drug delivery applications, it is very important that the collagen-fibril matrix is formulated in a way that exhibits tunability of one or more of its properties, including its microstructure, proteolytic biodegradability, mechanical properties, and its molecular transport properties. For this, we first need to understand the structure and self-assembly of collagen, which we look at in Chap. 2.

References

1. Y. Zhang, H.F. Chan, K.W. Leong, Advanced materials and processing for drug delivery: The past and the future. Adv. Drug Deliv. Rev. **65**, 104–120 (2013)
2. H. Hosseinkhani, M. Hosseinkhani, Biodegradable polymer-metal complexes for gene and drug delivery. Curr. Drug Saf. **4**, 79–83 (2009)
3. N.K. Mohtaram, A. Montgomery, S.M. Willerth, Biomaterial-based drug delivery systems for the controlled release of neurotrophic factors. Biomed. Mater. **8**, 022001 (2013)
4. B.P. Timko, D.S. Kohane. Drug-delivery systems for tunable and localized drug release. Isr. J. Chem. (2013) n/a-n/a
5. J. Robinson, in *Controlled Drug Delivery: Past Present and Future*, ed. by K. Park, (American Chemical Society, Washington, DC, 1997)
6. W.M. Saltzman, *Drug Delivery: Engineering Principles for Drug Therapy* (Oxford University Press, Oxford, 2001)
7. J.C. Sy, G. Seshadri, S.C. Yang, M. Brown, T. Oh, S. Dikalov, N. Murthy, M.E. Davis, Sustained release of a p38 inhibitor from non-inflammatory microspheres inhibits cardiac dysfunction. Nat. Mater. **7**, 863–868 (2008)
8. E. Choleris, S.R. Little, J.A. Mong, S.V. Puram, R. Langer, D.W. Pfaff, Microparticle-based delivery of oxytocin receptor antisense DNA in the medial amygdala blocks social recognition in female mice. Proc. Natl Acad. Sci. USA **104**, 4670–4675 (2007)
9. D.S. Kohane, G.L. Holmes, Y. Chau, D. Zurakowski, R. Langer, B.H. Cha, Effectiveness of muscimol-containing microparticles against pilocarpine-induced focal seizures. Epilepsia **43**, 1462–1468 (2002)
10. J.B. Ciolino, T.R. Hoare, N.G. Iwata, I. Behlau, C.H. Dohlman, R. Langer, D.S. Kohane, A drug-eluting contact lens. Invest. Ophthalmol. Vis. Sci. **50**, 3346–3352 (2009)
11. Y. Yeo, D.S. Kohane, Polymers in the prevention of peritoneal adhesions. Eur. J. Pharm. Biopharm. **68**, 57–66 (2008)
12. R. Langer, D.S.T. Hsieh, W. Rhine, J. Folkman, Control of release kinetics of macromolecules from polymers. J. Membr. Sci. **7**, 333–350 (1980)
13. R. Langer, T.H. Dean, S.L. Brown, Polymeric delivery systems for macromolecules, in *Biological Activities of Polymers*, (American Chemical Society, Washington, DC, 1982), pp. 95–105
14. A. Atala, Tissue engineering and regenerative medicine: Concepts for clinical application. Rejuvenation Res. **7**, 15–31 (2004)
15. J. Andrejecsk, W. Chang, J. Pober, W.M. Saltzman, Controlled protein delivery in the generation of microvascular networks. Drug Deliv. Transl. Res. **5**, 75–88 (2015)

16. D.I. Zeugolis, M. Raghunath, 2.215 – Collagen: Materials analysis and implant uses, in *Comprehensive Biomaterials*, ed. by D. Editor-in-Chief: Paul, (Elsevier, Oxford, 2011), pp. 261–278

17. K. Gelse, E. Poschl, T. Aigner, Collagens – Structure, function, and biosynthesis. Adv. Drug Deliv. Rev. **55**, 1531–1546 (2003)

18. A.K. Piez, Collagen, in *Encyclopedia of Polymer Science and Engineering*, ed. by M.H. F, (Wiley, New York, 1985), pp. 699–727

19. R. Parenteau-Bareil, R. Gauvin, F. Berthod, Collagen-based biomaterials for tissue engineering applications. Materials **3**, 1863–1887 (2010)

20. W. Friess, Collagen – Biomaterial for drug delivery. Eur. J. Pharm. Biopharm. **45**, 113–136 (1998)

21. M. Elder, *Markets for Advanced Wound Management Technologies* (BCC Research, Wellesley, 2014)

22. S.H. De Paoli Lacerda, B. Ingber, N. Rosenzweig, Structure-release rate correlation in collagen gels containing fluorescent drug analog. Biomaterials **26**, 7164–7172 (2005)

23. P.B. Malafaya, G.A. Silva, R.L. Reis, Natural-origin polymers as carriers and scaffolds for biomolecules and cell delivery in tissue engineering applications. Adv. Drug Deliv. Rev. **59**, 207–233 (2007)

24. L. Ma, C. Gao, Z. Mao, J. Zhou, J. Shen, Enhanced biological stability of collagen porous scaffolds by using amino acids as novel cross-linking bridges. Biomaterials **25**, 2997–3004 (2004)

25. R.J. Ruderman, C.W. Wade, W.D. Shepard, F. Leonard, Scanning electron microscopy of surfaces indicing shear hemolysis. J. Biomed. Mater. Res. **7**, 253–262 (1973)

26. Q. Jiang, N. Reddy, S. Zhang, N. Roscioli, Y. Yang, Water-stable electrospun collagen fibers from a non-toxic solvent and crosslinking system. J. Biomed. Mater. Res. A **101**(5), 1237–1247 (2012)

27. D.T. Cheung, M.E. Nimni, Mechanism of crosslinking of proteins by glutaraldehyde II. Reaction with monomeric and polymeric collagen. Connect. Tissue Res. **10**, 201–216 (1982)

28. J.M. Lee, H.H.L. Edwards, C.A. Pereira, S.I. Samii, Crosslinking of tissue-derived biomaterials in 1-ethyl-3-(3-dimethylaminopropyl)-carbodiimide (EDC). J. Mater. Sci. Mater. Med. **7**, 531–541 (1996)

29. J.M. Lee, C.A. Pereira, L.W.K. Kan, Effect of molecular-structure of poly (glycidyl ether) reagents on cross-linking and mechanical-properties of bovine pericardial xenograft materials. J. Biomed. Mater. Res. **28**, 981–992 (1994)

30. R. Tu, S.H. Shen, D. Lin, C. Hata, K. Thyagarajan, Y. Noishiki, R.C. Quijano, Fixation of bioprosthetic tissues with monofunctional and multifunctional polyepoxy compounds. J. Biomed. Mater. Res. **28**, 677–684 (1994)

31. L.H.H. Olde Damink, P.J. Dijkstra, M.J.A. van Luyn, P.B. van Wachem, P. Nieuwenhuis, J. Feijen, Cross-linking of dermal sheep collagen using a water-soluble carbodiimide. Biomaterials **17**, 765–773 (1996)

32. K.P. Rao, Recent developments of collagen-based materials for medical applications and drug delivery systems. J. Biomater. Sci. Polym. Ed. **7**, 623–645 (1995)

33. E. Khor, Methods for the treatment of collagenous tissues for bioprostheses. Biomaterials **18**, 95–105 (1997)

34. M.E. Nimni, D. Myers, D. Ertl, B. Han, Factors which affect the calcification of tissue-derived bioprostheses. J. Biomed. Mater. Res. **35**, 531–537 (1997)

35. L.L. Huang-Lee, D.T. Cheung, M.E. Nimni, Biochemical changes and cytotoxicity associated with the degradation of polymeric glutaraldehyde derived crosslinks. J. Biomed. Mater. Res. **24**, 1185–1201 (1990)

36. P.B. van Wachem, M.J. van Luyn, L.H. Olde Damink, P.J. Dijkstra, J. Feijen, P. Nieuwenhuis, Biocompatibility and tissue regenerating capacity of crosslinked dermal sheep collagen. J. Biomed. Mater. Res. **28**, 353–363 (1994)

37. S.A. Sell, P.S. Wolfe, K. Garg, J.M. McCool, I.A. Rodriguez, G.L. Bowlin, The use of natu-
    ral polymers in tissue engineering: A focus on electrospun extracellular matrix analogues.
    Polymers **2**, 522–553 (2010)
38. K.S. Weadock, E.J. Miller, L.D. Bellincampi, J.P. Zawadsky, M.G. Dunn, Physical crosslink-
    ing of collagen fibers: Comparison of ultraviolet irradiation and dehydrothermal treatment.
    J. Biomed. Mater. Res. **29**, 1373–1379 (1995)
39. D.G. Wallace, J. Rosenblatt, Collagen gel systems for sustained delivery and tissue engineer-
    ing. Adv. Drug Deliv. Rev. **55**, 1631–1649 (2003)
40. W. Friess, Collagen – Biomaterial for drug delivery1. Eur. J. Pharm. Biopharm. **45**, 113–136
    (1998)
41. Z. Ruszczak, W. Friess, Collagen as a carrier for on-site delivery of antibacterial drugs. Adv.
    Drug Deliv. Rev. **55**, 1679–1698 (2003)

Chapter 2
From *In Vivo* Synthesis to *In Vitro* Drug Delivery Device Formation

2.1 Collagen Biosynthesis

Type I collagen constitutes a major structural and mechanical component of connective tissues and organs. The biophysical properties of collagen that we touched upon in Chap. 1 are summed up in Fig. 2.1. To reiterate, what draws researchers to applying collagen for drug delivery is: (i) its mechanical strength that supports cell infiltration, proliferation, and movement in the native tissue, (ii) its ability to polymerize and form a porous fibril-matrix that can encapsulate drug molecule of interest to target a local site, (iii) its inherent cell-signaling potential which is facilitated by adhesion domains that engage in integrin-mediated cell binding [1], (iv) its ability to degrade into physiologically nontoxic products, and (v) its versatility and processability on an aqueous basis [2].

How does collagen self-assemble in vivo to form its complex, unique hierarchical structure that displays such an extraordinary set of properties? How does the collagen structure look like? Let us discuss these questions in this chapter.

The basic building block of this hierarchical structure is a collagen molecule consisting of three peptide chains (two α_1 (I) and one α_2 (I) chain). Collagen molecule comprises a central helical domain flanked on each end by non-helical telopeptide domains [3, 4] as shown in Fig. 2.2. The 300 nm long helical domain consists of Gly-X-Y repeats where the X and Y positions are often occupied by proline and hydroxyproline. These collagen molecules, also known as tropocollagen, are the fundamental building blocks of type I collagen. Tropocollagen molecules self-assemble in a hierarchical fashion as shown in Fig. 2.3 to form tissue-specific networks of fibrils that then combine to form suprafibrillar and tissue level structures [5, 6].

© The Author(s), under exclusive license to Springer Nature Switzerland AG 2021
R. Joshi, *Collagen Biografts for Tunable Drug Delivery*, SpringerBriefs in
Applied Sciences and Technology, https://doi.org/10.1007/978-3-030-63817-7_2

Fig. 2.1 Schematics showing biophysical properties of type I collagen that add to its advantages for forming a tissue engineering and drug delivery platform

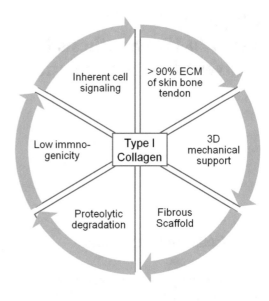

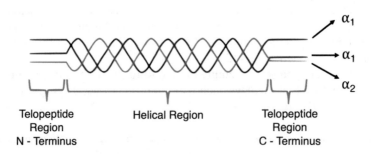

Fig. 2.2 Schematic of type I collagen molecular structure. Figure shows three polypeptide chains intertwined to create a right-handed helical structure. The N- and C-termini of the molecular structure contain the non-helical telopeptide regions

Collagen biosynthesis has been studied in greatest detail with regard to the fibrillar collagens. Many reviews such as that by Fratzl [7] and Kyle [6] have described the in vivo biosynthesis of collagen. The in vivo biosynthesis of collagen is indicated in the Fig. 2.3. In brief, collagen biosynthesis involves ribosomal production of individual tropocollagen alpha (α) chains which contribute to the triple helix stabilization, followed by hydroxylation of specific proline and lysine residues, which aid in the molecular cross-linking of collagen [5]. Processed polypeptide chains then undergo trimerization to form heterotrimeric procollagen molecules consisting of two α_1 (I) and a single α_2 (I) chains. Upon extrusion into the extracellular space, both amino- and carboxy-terminal propeptides are cleaved enzymatically [3], thus rendering the resultant tropocollagen molecule capable of fibril formation [8]. As the prefibrillar aggregates of staggered tropocollagen molecules assemble, lysyl oxidase binds and catalyzes cross-link formation between prefibrillar aggregates of staggered collagen molecules (telocollagen) to create covalently cross-linked dimers or trimers (called oligomers) [9]. These different oligomer precursors direct the progressive molecular packing, fibril assembly, and suprafibrillar network formation that eventually gives rise to tissue-specific form and function [5, 10].

So far, we saw that all collagens have a unique triple-helical structure configuration of three polypeptide units (alpha-chains), and a proper alignment of the alpha chains in vivo requires a highly complex enzymatic and chemical interaction. As such, the preparation of the collagen by alternate methods may result in improperly aligned alpha chains and, supposedly increase the immunogenicity of the collagen. Collagen is high in glycine, L-alanine, L-proline, and 4-hydroxyproline amino acids, low in sulfur, and contains no L-tryptophan. When heated (e.g., above 60 °C), the helical structure of collagen is irreversibly denatured to single α chains with some β and γ bands (gelatin). Such gelatin formulations display very low shape stability, poor mechanical strength, and low elasticity [11]. Gelatin formulations may not suitable for drug delivery unless their mechanical strength is improved by chemical processing or covalent cross-linking [12], because in the absence of cross-linking, gelatinous materials will display rapid degradation. The degradation term here refers to change in chemical, physical, or molecular structure or appearance (i.e., gross morphology) of a material.

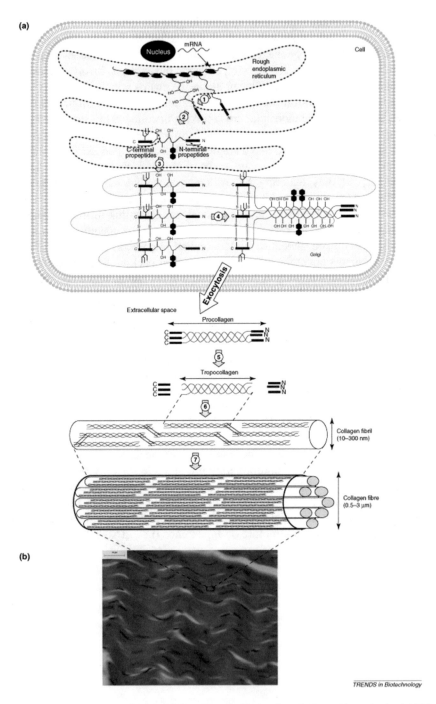

Fig. 2.3 Steps in collagen biosynthesis and assembly in vivo: Collagen biosynthesis. (**a**) This panel shows the major events in collagen fiber assembly. Modifications of collagen include hydrox-

2.2 Conventional Collagen Formulations

We previously saw in Chap. 1 that collagen-based drug delivery systems found in the market are scarce, considering the wealth of information that exists on collagen. The puzzle of why collagen-based products might not be reaching their full clinical potential can be solved by looking at how collagen formulations are conventionally prepared. Two main categories of collagen formulations prepared conventionally can be identified, as shown in Table 2.1. These are:

1. Non-dissociated Fibrillar Collagen – These formulations contain decellularized collagen ECM particulate matter, which is mechanically homogenized, acid-swollen, and finally lyophilized to form a sponge which may or may not be cross-linked. Such collagen formulations do not undergo polymerization since collagen fibers are never dissociated during this preparation method.
2. Soluble Collagen – These are obtained from pepsin or acid solubilization of mammalian tissues to form viscous collagen solutions, which are then lyophilized to form cross-linked or non-cross-linked sponge or injectable viscous gel. They exhibit fluid-like behavior under shear stress and become entangled again when the suspension is at rest.

The major limitation of these formulations is that they do not capture and capitalize on the inherent fibril self-assembly of collagen that occurs *in vivo*. As a result, the matrices formed from such formulations lack interfibrillar branching and simply represent entanglements of long individual fibrils, that lead to their poor shape definition, low mechanical integrity, poor handling, cell-induced contraction, and rapid proteolytic degradation [13–16]. To improve these properties, collagen materials are often subjected to exogenous cross-linking, achieved through chemical, enzymatic, or physical methods [13] but results in other unwanted problems as discussed below.

Chemical cross-linking of type I collagen matrices is typically performed using agents, such as glyoxal, formaldehyde, methylene diphenyl diisocyanate, hexamethylene diisocyanate, and most commonly glutaraldehyde [13]. Glutaraldehyde and formaldehyde treatment provides an advantage of cross-linking dry collagen material with a reagent in the vapor phase instead of treatment in the liquid phase [17]. Although these agents achieve the goal of cross-linking, they also exhibit

---

Fig 2.3 (continued) ylation of prolyl and lysyl residues (1), the addition of N-linked oligosaccharides (1), and glycosylation of hydroxylysyl residues in the endoplasmic reticulum (2), before chain alignment and disulfide bond formation (3), which results in the formation of the procollagen triple helix in the Golgi (4). After export from the cell, the N- and C-terminal propeptides are cleaved (5) and the resulting tropocollagen undergoes extensive cross-linking and self-assembly into collagen fibrils of diameters between 10 and 300 nm (6). These fibers in turn further assemble into larger fibers (0.5–3 µm diameter) (7). (**b**) Histological section of porcine patellar tendon showing a large number of intertwined collagen fibers with ligament fibroblasts, which are stained with hematoxylin (stains nuclei in blue-purple) and eosin (stains collagen fibers in pink). The scale bar represents 25 µm. (Figure adapted from work by Stuart Kyle [6], https://doi.org/10.1016/j.tibtech.2009.04.002 under the creative commons license https://creativecommons.org/licenses/by/4.0/)

Table 2.1 Major collagen formulations used in commercial drug delivery applications of collagen-based biomaterials

Collagen formulation	Preparation
Non dissociated, Fibrillar Collagen	Collagen Fiber → Homogenized Slurry → Acid Swollen → Lyophilized Sponge
Solubilized Collagen	Acid-Solubilized Collagen Monomer → Lyophilized Sponge / Viscous Injectable Gel

detrimental effects on cells and tissues [18] such as cytotoxicity [19, 20] or tissue calcification [21–23]. For example, depolymerization of polymeric glutaraldehyde cross-links has been reported to releases highly cytotoxic glutaraldehyde and mono-mer fragments into the recipient [21, 24–26]. Cross-linking with other chemical agents, for example, diphenyl phosphorylazide, 1-ethyl-3-(3-dimethylaminopropyl) carbodiimide (EDC)/N-hydroxysuccinimide (NHS), and oxygen species were proved to be nontoxic, but the cross-linked fibers were unstable in water and col-lapsed into films in aqueous or high humidity environments [27]. Besides, cross-linking can reduce porosity [23], limiting nutrient transport to cells.

Researchers have also attempted to use physical cross-linking techniques such as photooxidation, dehydrothermal treatments (DHT), and ultraviolet irradiation with photosensitizers (e.g., riboflavin) to avoid introducing potentially cytotoxic chemi-cal residuals into the system and retain the biocompatibility of collagen materials [13]. However, most of these physical treatments cannot yield a high enough cross-linking degree to meet the mechanical strength demand for drug delivery devices [16]. Furthermore, collagen is reported to have been partially denatured by these physical treatments [28]. Enzymatic cross-linking agents such as lysyl oxidase and tissue transglutaminase have also been used however limited due to feasibility issues [29] and concerns of apoptosis [30], respectively.

Thus, while materials formed without any cross-linking are characterized as mechanically unstable, too soft to handle, and unable to resist cell-induced contrac-tions, exogenous cross-linking has been shown to have detrimental effects on cells and tissues [18], such as cytotoxicity [19, 20] or tissue calcification [21–23] and partial denaturation of collagen itself [27, 28].

2.3 Collagen Building Blocks

Collagen extraction and formulation methods listed in Table 2.1 do not exploit the in vivo collagen synthesis and self-assembly imparting unique characteristics to the collagen matrix. However, porcine Type I collagen preserving the self-assembled collagen building blocks comprising monomers (telocollagens) and oligomers have been reported to be successfully acid solubilized and urified [31]. The structures of these isolated collagen building blocks viz. *oligomers*, and monomers such as *telo-collagens* and *atelocollagens* are shown in Fig. 2.4. The unique collagen building blocks extracted from porcine skin type I collagen (PSC) can predictably and repro-ducibly control the relevant fibril- and matrix-level properties such as matrix pore size, permeability and diffusivity, stiffness, and cell surface-receptor mediated sig-naling of collagen [5, 31, 32].

The collagen building blocks differ in their intermolecular cross-link content, composition, and cross-link chemistries [5, 31]. While the *oligomers* comprise small aggregates of collagen molecules (e.g., trimers), which retain collagen's tissue-specific, covalent intermolecular cross-links, *telocollagens* (monomers) are individ-ual collagen molecules, which lack these intermolecular covalent cross-links. The

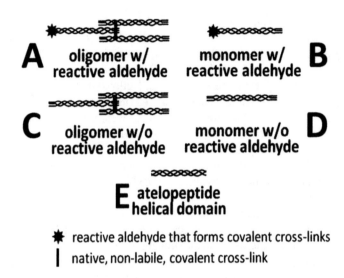

Fig. 2.4 A schematic of example soluble building blocks of a collagen-fibril polymer (**A** and **C**) Oligomer, (**B** and **D**) Telocollagen, and (**E**) Atelocollagen. Stars and bars represent reactive aldehydes and stable, mature covalent cross-links, respectively

telocollagen and oligomer formulations possess intact telopeptide regions and contain reactive aldehydes, building blocks generated from acid-labile, intermediate cross-links [5]. Upon in vivo polymerization, the process through which collagen fibrils assemble to form a gelled polymeric network, these reactive aldehydes spontaneously reform covalent, intermediate cross-links as part of the fibril-forming process.

Pepsin-solubilized (telopeptide-deficient) *atelocollagen* formulations are created when collagen is treated with proteolytic enzymes that remove the terminal telopeptide regions. As both the amino (N)- and carboxyl (C)-telopeptides play important roles in cross-linking and fibril formation, their complete removal results in an amorphous arrangement of collagen molecules and a consequent loss of the banded-fibril pattern in the reconstituted product [33].

The matrices formed from these building blocks have shown superior mechanobiological properties compared to conventional collagen formulations prepared under the same polymerization conditions [31]. The microstructural and mechanical properties given by these unique isolated building blocks are different from those obtained from collagen formulations in other categories. The relationship between matrix stiffness and fibril density, as exhibited by the building blocks was found to be important in regulating cell behavior and vessel morphogenesis [5]. Porcine Skin Collagen (PSC) showed decreased polymerization times, enhanced mechanical integrity, and a larger dynamic stiffness range than the other collagen formulations [5, 31]. Fundamental differences were shown between the PSC and conventional collagens on the molecular level, and these were attributed to the intermolecular cross-linking and ability of collagen to self-assemble as demonstrated by porcine skin collagen building blocks [5, 31, 34]. Inherent strengths of such collagen building-blocks should be capitalized for controlling the molecular release and formulating next-generation collagen-based drug delivery devices for local therapeutic delivery.

2.4 Drug Incorporation Method in Collagen Delivery Systems

It is important to note that the methods through which drugs are entrapped within or attached to collagen delivery systems play an important role in determining the efficacy of the drug delivery system [13, 35–37]. Since drug release can be influenced drastically by the approach taken to associate the drug with a collagen matrix, it is important to understand the current strategies of attaching the drug to collagen. An informed decision made in the selection of a strategy for drug attachment to collagen will endow us with a much better tool to engineer a drug delivery system with improved tunability and cell-instructive capacity.

Current strategies of drug incorporation can be separated into three distinct strategies (Fig. 2.5): (i) physical entrapment of drug, (ii) affinity binding based drug retention, and (iii) chemical or covalent immobilization of the drug into the collagen matrix. Physical admixing involves direct entrapment of a drug within the matrix and relies on diffusion to facilitate drug release into the surrounding tissue. Chemical immobilization usually involves covalent binding through the use of chemical cross-linkers and the drug is primarily released through degradation. Affinity binding involves affinity-based interaction between the drug and collagen substrate and the drug might be released by both diffusion and degradation. Let us review these techniques in a little bit of detail now.

(i) *Physical admixing/entrapment/adsorption of a drug into collagen:* Physical admixing involves dissolving or suspending the drug within a polymer reservoir to form a porous device. It is the most common strategy used for local drug delivery due to its simplicity and cost-effectiveness [38]. The rate of drug

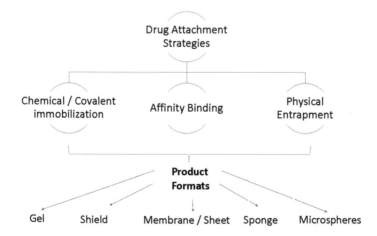

Fig. 2.5 Schematics of strategies of drug incorporation in collagen-based drug delivery systems

release is controlled by diffusion dominated mechanisms observed initially, followed by the further release as the reservoir degrades by surface or bulk erosion [39]. Fabrication methods for entrapping drugs involve lyophilization (freeze-drying), particulate leaching, phase emulsion (microspheres), and in situ polymerization (gels). Many of these methods start with slurries of shredded collagen or whole collagen tissue fragments that are exogenously cross-linked and combined with a drug at certain ratios before subjecting to lyophilization [40]. Sometimes, the drug is added after lyophilization as in cases of collagen sponge. During lyophilization, the pore-sizes that are formed within the matrices are often bigger than the size of the drug, resulting in diffusion dominated release. However, there is very little control over pore size during lyophilization, which limits the ability to tune drug entrapment and release from collagen. Moreover, harsh conditions of processing in the physical entrapment method (e.g., homogenization used during the emulsion method of formulating microspheres) can affect the bioactivity of encapsulated molecules by inactivating active sites or denaturing the drug [41].

(ii) *Affinity-based immobilization of drugs into collagen:* Rather than simply admixing a drug in collagen, site-specific tethering of drug to the collagen gives an option of extending drug release by modifying the interaction between the drug and matrix. Affinity can be described as the tendency for one molecule to bind to another. Affinity-based systems utilize the molecular interaction between the therapeutic agent and the delivery vehicle. The strength of the interaction depends on the variety of molecular forces: charge, hydrophobicity, hydrogen bonding, and Van der Waals forces [42]. This non-covalent physical adsorption technique involves adsorption of drugs onto surfaces typically exploiting direct charge–charge or secondary drug-matrix affinity interactions, or indirect interaction via intermediate proteins or other biological molecules (e.g., heparin [43], fibronectin [44]). Such interactions have been employed to deliver basic fibroblast growth factor (bFGF) and vascular endothelial growth factor (VEGF) through engineered biomimetic collagen matrices, showing controlled diffusion and matrix degradation to induce angiogenesis [43, 45–48].

(iii) *Chemical/covalent immobilization of drug on collagen:* Immobilization of the drug within the collagen matrix can also be achieved by its covalent conjugation to collagen. Covalent binding of drugs to collagen matrices can sustain drug release for a longer period and offer control over the amount and spatial distribution of a drug in collagen matrices. Drugs can be conjugated to collagen matrices via functional groups, which are incorporated by copolymerization or through a chemical treatment. For example, to overcome rapid diffusion and clearance from the implant site and to increase its conformational stability, recombinant transforming growth factor β2 (TGF-β2) was covalently bound to injectable bovine dermal fibrillar collagen (FC), using a difunctional polyethylene glycol (PEG) linker to create FC-PEG-TGF-β2 sequences throughout the matrix [49]. The activity of the covalently bound TGF-β2 was compared to admixed TGF-β2, and results showed that covalent binding of TGF-β2 to collagen resulted in a significantly larger and longer-lasting TGF-β2 response than that observed with admixed formulations of collagen and TGF-β2. It should be noted, however, that despite the advantages offered by this method of drug incorporation, it is difficult to selectively assign specificity of the coupling site on the conjugated drug as binding interactions are specific to each drug and difficult to predict. Also, biomolecules may lose their bioactivity if screening or damage of the active pockets occurs during the immobilization [50].

The ultimate success of any of the above methods of drug loading, whether physical entrapment, affinity-based retention, or covalent immobilization, is dependent on the preservation of collagen's native physiological properties. Physical entrapment and affinity-based molecular retention methods are often confounded by the weak mechanical properties of conventional collagen formulations. As mentioned earlier, materials formed without any cross-linking are characterized as mechanically unstable, too soft to handle, and unable to resist cell-induced contractions [13–16] thus failing to support cell ingrowth and migration required for tissue regeneration. On the other hand, exogenous cross-linking [13, 27, 51–55] or chemical immobilization-based approaches can lead to detrimental effects on cells and tissues [18], such as cytotoxicity [19, 20] or tissue calcification [21–23] and partial denaturation of collagen itself [27, 28]. Consequently, current collagen-based products suffer from problems related to mechanical integrity, inability to give controlled release, and inflammation-based tissue response. This limits the clinical success of collagen for tissue engineering and molecular delivery applications [13, 37].

2.5   Drug Delivery from Conventional Collagen Formulations: State-of-the-Art and Current Limitations

Interest has been rising in combining growth factors into collagen [50, 56, 57] since many growth factors have been recently recognized as key signaling molecules, especially those helping in wound healing [58]. For example, platelet-derived growth factor-BB (PDGF-BB) is important for the granulation tissue formation and stem cell recruitment, vascular endothelial growth factor-A (VEGF-A) is needed to induce blood vessel growth for sustaining the granulation tissue, and fibroblast growth factors (FGFs) are crucial for both wound reepithelization and angiogenesis [59]. However, none of these growth factor-loaded collagen matrices have been clinically viable. The reasons are their rapid clearance from the matrix and/or degradation of soluble growth factors at the implant site [60, 61]. This raises an important question on the ability of collagen to serve as a matrix to achieve a controlled release of growth factors.

Interestingly, to date, numerous studies report applications of collagen for controlled drug delivery as ophthalmological shields, antibiotic-loaded sponges, drug-loaded microspheres, and injectable collagen gels, and have been extensively reviewed previously [13, 35–37]. Selective examples are provided in Table 2.2. However, a close inspection of the current collagen market shows that only a few

Table 2.2 Examples of type I collagen-based drug delivery in research

Molecules delivered	*In vitro/in vivo* applications	Drug-binding approach
1. *Gels*		
Pilocarpine [71] TGF-β3 [72] Doxorubicin [73] Ketorolac [74]	Ophthalmic treatment Craniosynostosis treatment Cancer chemotherapy Treating inflammation	*Physical:* Direct admixing in collagen solution, then allowing collagen to polymerize into a gel by incubation at 37 °C
Transforming growth factor TGF-β2 [49]	Facilitating tissue repair	*Chemical:* Covalent binding to collagen through difunctional PEG
Thrombin resistant mutant of Fibroblast growth factor (FGF)-1 (R136K) with a collagen binding domain (CBD) [75]	Smooth muscle cell proliferation	Chemical: Chimeric collagen-binding domain-based attachment
2. *Shields*		
Plasmid DNA [76]	Gene therapy for healing after glaucoma surgery	*Physical:* Plasmid absorbed into the collagen shield
Gentamicin GA and Vancomycin VA [77] Tobramycin, Pilocarpine [78]	Antibiotic therapy through collagen contact lenses Keratoplasty treatment	*Physical:* Presoaking collagen shield in drug solution right before application
3. *Membrane/sheet*		
Vascular endothelial growth factor VEGF [79], Nifedipine [80]	Useful for paracrine assays and angiogenesis Transdermal delivery devices for wound dressings	*Physical*: Admixing in collagen gel followed by vitrification *Physical*: Mixing in alginate and then into collagen membrane
4. *Microspheres/nanoparticles*		
Retinol, Tretinoin, or Tetracaine, and Lidocaine in free base form [81]	Carriers for lipophilic drugs	*Physical:* Drug encapsulated by emulsion into cross-linked collagen microspheres
Cyclosporines [82]	Delivery to the ocular surface to prevent corneal graft rejection	*Physical:* Collagen-particles encapsulating cyclosporine suspended in methylcellulose
5. *Sponge*		
Retinoic acid RA [83]	Endothelial regeneration in prosthetic bypass grafts	*Chemical:* Chimeric domain binding to sponge
5-Fluorouracil [84] Gentamicin [85, 86] rhBMP-2 [87]	Reduces intraocular pressure Wound healing Bone remodeling	*Physical:* Lyophilized sponge rehydrated and soaked in drug/growth factor solution

collagen-based drug delivery formulations have made it into clinical trials or are currently marketed [13, 62, 63]. Despite the numerous advantages and wide research on applications collagen as an excellent natural biomaterial [3, 7, 13, 35, 62], its use as a vehicle for controlling local growth factor release is seen to be limited [37, 64]. Perhaps matrices formed using isolated unique collagen building blocks described in Sect. 2.2 can overcome the inability to achieve controlled release of biomolecules. Such an application of collagen-building blocks for drug delivery is shown by Joshi [65].

2.6 Drug Delivery Through Various Formats of Collagen

The versatility of collagen lends itself well to a variety of medical applications including but not limited to wound care, oral surgery, cardiovascular systems, neurology, urology, and orthopedics. The formats of collagen used in these applications are many, and selective examples are listed in Table 2.2.

To get a glimpse of how some of these formats of collagen might look like under a scanning electron microscopic, please refer to Fig. 2.6 adapted from Meyer's work [66]. Figure 2.6a–f shows the appearance of a selected collagenous materials manufactured by variants of the processing sequences. Structures of the freeze-dried raw tissues, namely, decellularized pericardium (Fig. 2.6a), and hairless decellularized skin (Fig. 2.6d) show that while the pericardial fibrosa delaminates to some extent, the collagen fiber bundles clump in the decellularized skin. Catgut (Fig. 2.6f) is manufactured from purified SIS which is cut in small strands, twisted, and convection dried under tension. Porous sponges achieved by freeze-drying are shown in Fig. 2.6b, while compact films achieved by convection drying are shown in Fig. 2.6d. While gelatin (Fig. 2.6c) is a denatured collagen, microscopically no difference is observed between gelatin and the film with the triple helices still present in it.

Sponges: Collagen sponges were originally developed as wound dressings due to their ability to absorb large quantities of tissue exudates, adherence to wet wound bed with preservation of the low-moist environment, and shielding against mechani-

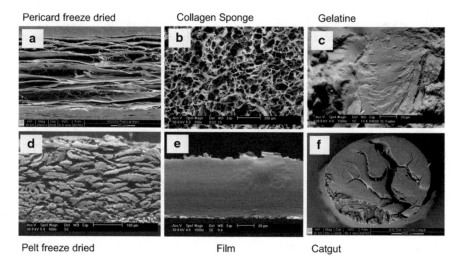

Fig. 2.6 Comparison of microscopic structures of principally different processed materials. Pericardium (**a**) and skin (**d**) are decellularized and freeze-dried, for sponge (**b**) purified skin is minced and freeze-dried, for films convection-dried (**e**), instead. Catgut (**f**) consists of purified small intestine (SIS) which is cut in strands, twisted, stabilized by cross-linking, and convection-dried. Gelatin (**c**) is hydrolyzed collagen of skin or bone which have been convection-dried. (Figure is adapted from work by Meyer [66], https://doi.org/10.1186/s12938-019-0647-0, under the creative commons license https://creativecommons.org/licenses/by/4.0/)

cal harm and secondary bacterial infection [67]. Growth factors have been coated on collagen sponge to give recovery from dermal and epidermal wounds [68, 69]. Collagen sponges are generally prepared by lyophilizing aqueous collagen preparations [13] that yield collagen sponges with high porosity and fibril interconnectivity. The porosity of the lyophilized collagen can be altered by varying the collagen concentration and the freezing rate, which allows for some degree of control over the design of the sponge [70].

Another method of loading lyophilized sponge, apart from coating them with drug solution, is to soak the sponge in aqueous drug formulations before implantation. For example, porous collagen sponges have been soaked in antibiotic solutions (e.g., gentamicin) [85, 86] and in growth factor solutions (e.g., rhBMP) for delivery to a tissue of interest [87]. Besides, collagen can be combined with other materials such as elastin [88], fibronectin, and hyaluronate [89] or glycosaminoglycans [90, 91] to aid in the delivery of drugs which do not interact well with collagen. The starting collagen material can be cross-linked with agents like glutaraldehyde and dehydrothermal treatments (DHT) to achieve highly resilient materials [90, 91]. However, the use of such cross-linking agents is not always effective as discussed in Sect. 2.6. Sponges also suffer from the problem of releasing the entrapped factors quickly [92] giving a burst release profile in most cases [13].

Gels: Collagen gels are primarily used in aqueous injectable systems that are initially liquid but solidify after administration to the tissue. In situ polymerization methods offer an advantage of injectability and spatial control with better mechanical properties over other collagen-based devices such as implantable collagen sponges or sheets. For most gel formulations, the drug is admixed or physically entrapped with collagen in liquid form at a certain ratio, and then allowed to gel when the temperature is raised to 37 °C (body temperature), as is the case with drugs such as pilocarpine, TGF-β, doxorubicin, and ketorolac [71–74]. Although such collagen gel systems show promise in drug delivery, their open pore structure causes diffusion-dominated release, which is undesirable due to little or no control over drug release rates.

Shields: Collagen shields have been primarily used as therapeutic devices for ophthalmological conditions such as plasmid delivery for glaucoma treatment or contact lenses to promote corneal epithelial healing and deliver hydro soluble drugs [76]. Shields typically start in a dehydrated form and have to be soaked with drugs in liquid solution before application. The thin collagen films conform to the shape of the cornea when applied to the eye and can provide sufficient oxygen transmission, as well as act as short-term bandage lenses [93]. As the shields dissolve, they provide a layer of collagen solution that lubricates the surface of the eye, minimizes rubbing of the lids on the cornea, and fosters epithelial healing [35]. However, some disadvantages still limit the application of collagen shields such as incomplete transparency, slight discomfort, complex insertion technique, and a short period of working before dissolution. For mechanical strength imparting reason, cross-linking is performed on shields already loaded with drugs but that endangers the chemical integrity of the active substance [94].

Microspheres: Collagen microparticulate systems have been used for encapsulating several antibiotics, steroids, growth factors, and hydrophilic and hydrophobic drugs for therapeutic purposes due to their small particle size, large surface area, and ability to disperse in water to form colloidal solutions [35]. Microspheres can provide regulation of release by controlling the shell material and protection of drug until its delivery is needed [95]. Moreover, microspheres can create gradients in the concentration of growth factors that can direct cell migration, create patterns of cell differentiation, and direct tissue organization into complex structures such as branching networks of vascular systems [2].

Despite many successful studies on collagen microspheres, the transition to collagen as the primary biomaterial for microsphere technology is hindered by limitations in manufacturing material, methods, and use of solvents due to risks of collagen denaturation. Most of the methods of the formulation are tedious, requiring that each step (i.e., droplet generation, gelation, and extraction) be performed separately [96]. Furthermore, the microsphere has to be cross-linked exogenously in most cases, in order to avoid the possibility of losing the mechanical integrity and shape of the device, but that leads to detrimental effects of exogenous cross-links.

Summing up these cases, again, it is clear that there is a significant challenge in the design and manufacture of multifunctional biograft materials capable of providing tunable molecular delivery due to the poor mechanical properties, rapid proteolytic degradation, and inability of collagen formulations to demonstrate physiologically relevant self-assembly. However, the use of self-assembling collagen-fibril biograft materials in various drug delivery formats can be explored to overcome existing limitations. A successful design and development of self-assembling, multifunctional 3D collagen-fibril biograft materials can be achieved with a broad range of tunable physical and molecular delivery properties. More specifically, collagen polymers specified by their intermolecular cross-link composition and self-assembly capacity can be used to customize and design materials in terms of (1) collagen fibril microstructure and (2) proteolytic degradability, which we discuss in the next chapter.

References

1. J. Jokinen, E. Dadu, P. Nykvist, J. Käpylä, D.J. White, J. Ivaska, P. Vehviläinen, H. Reunanen, H. Larjava, L. Häkkinen, J. Heino, Integrin-mediated cell adhesion to type I collagen fibrils. J. Biol. Chem. **279**, 31956–31963 (2004)
2. D.I. Zeugolis, M. Raghunath, 2.215. Collagen, Materials analysis and implant uses, in *Comprehensive Biomaterials*, ed. by D. Paul, (Elsevier, Oxford, 2011), pp. 261–278
3. K. Gelse, E. Poschl, T. Aigner, Collagens–structure, function, and biosynthesis. Adv. Drug Deliv. Rev. **55**, 1531–1546 (2003)
4. S. Ricard-Blum, F. Ruggiero, The collagen superfamily: From the extracellular matrix to the cell membrane. Pathologie-biologie **53**, 430–442 (2005)
5. J.L. Bailey, P.J. Critser, C. Whittington, J.L. Kuske, M.C. Yoder, S.L. Voytik-Harbin, Collagen oligomers modulate physical and biological properties of three-dimensional self-assembled matrices. Biopolymers **95**, 77–93 (2011)
6. S. Kyle, A. Aggeli, E. Ingham, M.J. McPherson, Production of self-assembling biomaterials for tissue engineering. Trends Biotechnol. **27**, 423–433 (2009)

7. F. Peter, Collagen structure and mechanics, in *Collagen Structure and Mechanics: An Introduction*, ed. by F. Peter, (Springer Science + Business Media, LLC, New York, 2008)
8. B.d.C.J. Rossert, Type I collagen: Structure, synthesis and regulation, in *Principles in Bone Biology*, ed. by L. G. R. J. P. Bilezkian, G. A. Rodan, (Academic Press, Orlando, 2002), pp. 189–210
9. R.G. Paul, A.J. Bailey, Chemical stabilisation of collagen as a biomimetic. Sci. World J. **3**, 138–155 (2003)
10. W.J.-J. Eyre David, Collagen cross-links, in *Topics in Current Chemistry*, (Springer, Berlin/ Heidelberg, 2005), pp. 207–229
11. R. Dash, M. Foston, A.J. Ragauskas, Improving the mechanical and thermal properties of gelatin hydrogels cross-linked by cellulose nanowhiskers. Carbohydr. Polym. **91**, 638–645 (2013)
12. Q. Xing, K. Yates, C. Vogt, Z. Qian, M.C. Frost, F. Zhao, Increasing mechanical strength of gelatin hydrogels by divalent metal ion removal. Sci. Rep. **4**, 4706 (2014)
13. W. Friess, Collagen–biomaterial for drug delivery. Eur. J. Pharm. Biopharm. **45**, 113–136 (1998)
14. S.H. De Paoli Lacerda, B. Ingber, N. Rosenzweig, Structure-release rate correlation in collagen gels containing fluorescent drug analog. Biomaterials **26**, 7164–7172 (2005)
15. P.B. Malafaya, G.A. Silva, R.L. Reis, Natural-origin polymers as carriers and scaffolds for biomolecules and cell delivery in tissue engineering applications. Adv. Drug Deliv. Rev. **59**, 207–233 (2007)
16. L. Ma, C. Gao, Z. Mao, J. Zhou, J. Shen, Enhanced biological stability of collagen porous scaffolds by using amino acids as novel cross-linking bridges. Biomaterials **25**, 2997–3004 (2004)
17. R.J. Ruderman, C.W. Wade, W.D. Shepard, F. Leonard, Scanning electron microscopy of surfaces indicing shear hemolysis. J. Biomed. Mater. Res. **7**, 253–262 (1973)
18. K.P. Rao, Recent developments of collagen-based materials for medical applications and drug delivery systems. J. Biomater. Sci. Polym. Ed. **7**, 623–645 (1995)
19. E. Khor, Methods for the treatment of collagenous tissues for bioprostheses. Biomaterials **18**, 95–105 (1997)
20. M.E. Nimni, D. Myers, D. Ertl, B. Han, Factors which affect the calcification of tissue-derived bioprostheses. J. Biomed. Mater. Res. **35**, 531–537 (1997)
21. L.L. Huang-Lee, D.T. Cheung, M.E. Nimni, Biochemical changes and cytotoxicity associated with the degradation of polymeric glutaraldehyde derived crosslinks. J. Biomed. Mater. Res. **24**, 1185–1201 (1990)
22. P.B. van Wachem, M.J. van Luyn, L.H. Olde Damink, P.J. Dijkstra, J. Feijen, P. Nieuwenhuis, Biocompatibility and tissue regenerating capacity of crosslinked dermal sheep collagen. J. Biomed. Mater. Res. **28**, 353–363 (1994)
23. S.A. Sell, P.S. Wolfe, K. Garg, J.M. McCool, I.A. Rodriguez, G.L. Bowlin, The use of natural polymers in tissue engineering: A focus on electrospun extracellular matrix analogues. Polymers **2**, 522–553 (2010)
24. J.K. Law, J.R. Parsons, F.H. Silver, A.B. Weiss, An evaluation of purified reconstituted type 1 collagen fibers. J. Biomed. Mater. Res. **23**, 961–977 (1989)
25. M.J. Morykwas, In vitro properties of crosslinked, reconstituted collagen sheets. J. Biomed. Mater. Res. **24**, 1105–1110 (1990)
26. A. Simionescu, D. Simionescu, R. Deac, Lysine-enhanced glutaraldehyde crosslinking of collagenous biomaterials. J. Biomed. Mater. Res. **25**, 1495–1505 (1991)
27. Q. Jiang, N. Reddy, S. Zhang, N. Roscioli, Y. Yang, Water-stable electrospun collagen fibers from a non-toxic solvent and crosslinking system. J. Biomed. Mater. Res. A (2012)
28. K.S. Weadock, E.J. Miller, L.D. Bellincampi, J.P. Zawadsky, M.G. Dunn, Physical crosslinking of collagen fibers: Comparison of ultraviolet irradiation and dehydrothermal treatment. J. Biomed. Mater. Res. **29**, 1373–1379 (1995)
29. J.D. Berglund, M.M. Mohseni, R.M. Nerem, A. Sambanis, A biological hybrid model for collagen-based tissue engineered vascular constructs. Biomaterials **24**, 1241–1254 (2003)

30. W.M. Elbjeirami, E.O. Yonter, B.C. Starcher, J.L. West, Enhancing mechanical properties of tissue-engineered constructs via lysyl oxidase crosslinking activity. J. Biomed. Mater. Res. A **66**, 513–521 (2003)
31. S.T. Kreger, B.J. Bell, J. Bailey, E. Stites, J. Kuske, B. Waisner, S.L. Voytik-Harbin, Polymerization and matrix physical properties as important design considerations for soluble collagen formulations. Biopolymers **93**, 690–707 (2010)
32. C.F. Whittington, E. Brandner, K.Y. Teo, B. Han, E. Nauman, S.L. Voytik-Harbin, Oligomers modulate interfibril branching and mass transport properties of collagen matrices. Microsc Microanal, 1–11 (2013)
33. K.E. Kadler, D.F. Holmes, J.A. Trotter, J.A. Chapman, Collagen fibril formation. Biochem. J. **316**(Pt 1), 1–11 (1996)
34. C.F. Whittington, M.C. Yoder, S.L. Voytik-Harbin, Collagen-polymer guidance of vessel network formation and stabilization by endothelial colony forming cells in vitro. Macromol. Biosci. **13** (2013).
35. C.H. Lee, A. Singla, Y. Lee, Biomedical applications of collagen. Int. J. Pharm. **221**, 1–22 (2001)
36. Z. Ruszczak, W. Friess, Collagen as a carrier for on-site delivery of antibacterial drugs. Adv. Drug Deliv. Rev. **55**, 1679–1698 (2003)
37. D.G. Wallace, J. Rosenblatt, Collagen gel systems for sustained delivery and tissue engineering. Adv. Drug Deliv. Rev. **55**, 1631–1649 (2003)
38. M. Ehrbar, V.G. Djonov, C. Schnell, S.A. Tschanz, G. Martiny-Baron, U. Schenk, J. Wood, P.H. Burri, J.A. Hubbell, A.H. Zisch, Cell-demanded liberation of VEGF121 from fibrin implants induces local and controlled blood vessel growth. Circ. Res. **94**, 1124–1132 (2004)
39. N.K. Mohtaram, A. Montgomery, S.M. Willerth, Biomaterial-based drug delivery systems for the controlled release of neurotrophic factors. Biomed. Mater. **8**, 022001 (2013)
40. R.A. Brown, Direct collagen-material engineering for tissue fabrication. Tissue Eng. A **19**, 1495–1498 (2013)
41. R.R. Chen, D.J. Mooney, Polymeric growth factor delivery strategies for tissue engineering. Pharm. Res. **20**, 1103–1112 (2003)
42. N.X. Wang, H.A. von Recum, Affinity-based drug delivery. Macromol. Biosci. **11**, 321–332 (2011)
43. M. Markowicz, A. Heitland, G.C. Steffens, N. Pallua, Effects of modified collagen matrices on human umbilical vein endothelial cells. Int. J. Artificial Organs **28**, 1251–1258 (2005)
44. A.L. Weiner, S.S. Carpenter-Green, E.C. Soehngen, R.P. Lenk, M.C. Popescu, Liposome-collagen gel matrix: A novel sustained drug delivery system. J. Pharm. Sci. **74**, 922–925 (1985)
45. M.J. Wissink, R. Beernink, J.S. Pieper, A.A. Poot, G.H. Engbers, T. Beugeling, W.G. van Aken, J. Feijen, Immobilization of heparin to EDC/NHS-crosslinked collagen. Characterization and in vitro evaluation. Biomaterials **22**, 151–163 (2001)
46. M.J. Wissink, R. Beernink, A.A. Poot, G.H. Engbers, T. Beugeling, W.G. van Aken, J. Feijen, Improved endothelialization of vascular grafts by local release of growth factor from heparinized collagen matrices. J. Control. Release **64**, 103–114 (2000)
47. G.C. Steffens, C. Yao, P. Prevel, M. Markowicz, P. Schenck, E.M. Noah, N. Pallua, Modulation of angiogenic potential of collagen matrices by covalent incorporation of heparin and loading with vascular endothelial growth factor. Tissue Eng. **10**, 1502–1509 (2004)
48. G. Grieb, A. Groger, A. Piatkowski, M. Markowicz, G.C. Steffens, N. Pallua, Tissue substitutes with improved angiogenic capabilities: an in vitro investigation with endothelial cells and endothelial progenitor cells. Cells Tissues Organs **191**, 96–104 (2010)
49. H. Bentz, J.A. Schroeder, T.D. Estridge, Improved local delivery of TGF-beta2 by binding to injectable fibrillar collagen via difunctional polyethylene glycol. J. Biomed. Mater. Res. **39**, 539–548 (1998)
50. K. Lee, E.A. Silva, D.J. Mooney, Growth factor delivery-based tissue engineering: General approaches and a review of recent developments. J. R. Soc. Interface **8**, 153–170 (2011)

51. D.T. Cheung, M.E. Nimni, Mechanism of crosslinking of proteins by glutaraldehyde II. Reaction with monomeric and polymeric collagen. Connect. Tissue Res. **10**, 201–216 (1982)
52. J.M. Lee, H.H.L. Edwards, C.A. Pereira, S.I. Samii, Crosslinking of tissue-derived biomaterials in 1-ethyl-3-(3-dimethylaminopropyl)-carbodiimide (EDC). J. Mater. Sci. Mater. Med. **7**, 531–541 (1996)
53. J.M. Lee, C.A. Pereira, L.W.K. Kan, Effect of molecular-structure of Poly(Glycidyl Ether) reagents on cross-linking and mechanical-properties of bovine pericardial xenograft materials. J. Biomed. Mater. Res. **28**, 981–992 (1994)
54. R. Tu, S.H. Shen, D. Lin, C. Hata, K. Thyagarajan, Y. Noishiki, R.C. Quijano, Fixation of bioprosthetic tissues with monofunctional and multifunctional polyepoxy compounds. J. Biomed. Mater. Res. **28**, 677–684 (1994)
55. L.H.H. Olde Damink, P.J. Dijkstra, M.J.A. van Luyn, P.B. van Wachem, P. Nieuwenhuis, J. Feijen, Cross-linking of dermal sheep collagen using a water-soluble carbodiimide. Biomaterials **17**, 765–773 (1996)
56. J.S. Pieper, T. Hafmans, P.B. van Wachem, M.J.A. van Luyn, L.A. Brouwer, J.H. Veerkamp, T.H. van Kuppevelt, Loading of collagen-heparan sulfate matrices with bFGF promotes angiogenesis and tissue generation in rats. J. Biomed. Mater. Res. **62**, 185–194 (2002)
57. D.S. Thoma, N. Nänni, G.I. Benic, F.E. Weber, C.H.F. Hämmerle, R.E. Jung, Effect of platelet-derived growth factor-BB on tissue integration of cross-linked and non-cross-linked collagen matrices in a rat ectopic model. Clin. Oral Implants Res. **26**, 263–270 (2015)
58. T.N. Demidova-Rice, J.T. Durham, I.M. Herman, Wound healing angiogenesis: Innovations and challenges in acute and chronic wound healing. Adv. Wound Care **1**, 17–22 (2012)
59. B. Buchberger, M. Follmann, D. Freyer, H. Huppertz, A. Ehm, J. Wasem, The importance of growth factors for the treatment of chronic wounds in the case of diabetic foot ulcers. GMS Health Technol. Assess **6** (2010) Doc12
60. Y.M. Elçin, V. Dixit, G. Gitnick, Extensive in vivo angiogenesis following controlled release of human vascular endothelial cell growth factor: Implications for tissue engineering and wound healing. Artif. Organs **25**, 558–565 (2001)
61. S.T.M. Nillesen, P.J. Geutjes, R. Wismans, J. Schalkwijk, W.F. Daamen, T.H. van Kuppevelt, Increased angiogenesis and blood vessel maturation in acellular collagen–heparin scaffolds containing both FGF2 and VEGF. Biomaterials **28**, 1123–1131 (2007)
62. D. S., Global markets and technologies for advanced drug delivery systems, 153 (2014).
63. G. Kevin, Global markets for drug-device combinations, BCC Market Reserach Report, No. PHM045C, ISBN: 0-89336-243-3, (2013).
64. K. Fujioka, M. Maeda, T. Hojo, A. Sano, Protein release from collagen matrices. Adv. Drug Deliv. Rev. **31**, 247–266 (1998)
65. R. Joshi, Designer collagen-fibril biograft materials for tunable molecular delivery, Open Access Dissertations, in *Biomedical Engineering*, (Purddue University, West Lafayette, 2016)
66. M. Meyer, Processing of collagen based biomaterials and the resulting materials properties. Biomed. Eng. Online **18**, 24–24 (2019)
67. S. Suzuki, K. Matsuda, N. Isshiki, Y. Tamada, K. Yoshioka, Y. Ikada, Clinical evaluation of a new bilayer "artificial skin" composed of collagen sponge and silicone layer. Br. J. Plast. Surg. **43**, 47–54 (1990)
68. P.M. Royce, T. Kato, K. Ohsaki, A. Miura, The enhancement of cellular infiltration and vascularisation of a collagenous dermal implant in the rat by platelet-derived growth factor BB. J. Dermatol. Sci. **10**, 42–52 (1995)
69. M.G. Marks, C. Doillon, F.H. Silver, Effects of fibroblasts and basic fibroblast growth factor on facilitation of dermal wound healing by type I collagen matrices. J. Biomed. Mater. Res. **25**, 683–696 (1991)
70. J.F. Burke, I.V. Yannas, W.C. Quinby Jr., C.C. Bondoc, W.K. Jung, Successful use of a physiologically acceptable artificial skin in the treatment of extensive burn injury. Ann. Surg. **194**, 413–428 (1981)

71. A.L. Rubin, K.H. Stenzel, T. Miyata, M.J. White, M. Dunn, Collagen as a vehicle for drug delivery. Preliminary report. J. Clin. Pharmacol. **13**, 309–312 (1973)
72. S. Premaraj, B.L. Mundy, D. Morgan, P.L. Winnard, M.P. Mooney, A.M. Moursi, Sustained delivery of bioactive cytokine using a dense collagen gel vehicle collagen gel delivery of bioactive cytokine. Arch. Oral Biol. **51**, 325–333 (2006)
73. C. Kojima, T. Suehiro, K. Watanabe, M. Ogawa, A. Fukuhara, E. Nishisaka, A. Harada, K. Kono, T. Inui, Y. Magata, Doxorubicin-conjugated dendrimer/collagen hybrid gels for metastasis-associated drug delivery systems. Acta Biomater. **9**, 5673–5680 (2013)
74. S.E. Fu, C. R., Fleitman J. S., De Leung M. C., Collagen containing ophthalmic formulation, in: E. Patent (Ed.), 1990
75. Y. Pang, H.P. Greisler, Using a type 1 collagen-based system to understand cell-scaffold interactions and to deliver chimeric collagen-binding growth factors for vascular tissue engineering. J. Investig. Med. **58**, 845–848 (2010)
76. G.J. Angella, M.B. Sherwood, L. Balasubramanian, J.W. Doyle, M.F. Smith, G. van Setten, M. Goldstein, G.S. Schultz, Enhanced short-term plasmid transfection of filtration surgery tissues. Invest. Ophthalmol. Vis. Sci. **41**, 4158–4162 (2000)
77. R.B. Phinney, S.D. Schwartz, D.A. Lee, B.J. Mondino, Collagen-shield delivery of gentamicin and vancomycin. Arch. Ophthalmol. **106**, 1599–1604 (1988)
78. J.V. Aquavella, J.J. Ruffini, J.A. LoCascio, Use of collagen shields as a surgical adjunct. J Cataract Refract Surg **14**, 492–495 (1988)
79. T. Takezawa, T. Takeuchi, A. Nitani, Y. Takayama, M. Kino-Oka, M. Taya, S. Enosawa, Collagen vitrigel membrane useful for paracrine assays in vitro and drug delivery systems in vivo. J. Biotechnol. **131**, 76–83 (2007)
80. D. Thacharodi, K.P. Rao, Rate-controlling biopolymer membranes as transdermal delivery systems for nifedipine: Development and in vitro evaluations. Biomaterials **17**, 1307–1311 (1996)
81. B. Rossler, J. Kreuter, D. Scherer, Collagen microparticles: Preparation and properties. J. Microencapsul. **12**, 49–57 (1995)
82. B.M. Gebhardt, H.E. Kaufman, Collagen as a delivery system for hydrophobic drugs: Studies with cyclosporine. J. Ocul. Pharmacol. Ther. **11**, 319–327 (1995)
83. L.P. Brewster, C. Washington, E.M. Brey, A. Gassman, A. Subramanian, J. Calceterra, W. Wolf, C.L. Hall, W.H. Velander, W.H. Burgess, H.P. Greisler, Construction and characterization of a thrombin-resistant designer FGF-based collagen binding domain angiogen. Biomaterials **29**, 327–336 (2008)
84. J.S. Kay, B.S. Litin, M.A. Jones, A.W. Fryczkowski, M. Chvapil, J. Herschler, Delivery of antifibroblast agents as adjuncts to filtration surgery–Part II: Delivery of 5-fluorouracil and bleomycin in a collagen implant: pilot study in the rabbit. Ophthalmic Surg. **17**, 796–801 (1986)
85. H.J. Rutten, P.H. Nijhuis, Prevention of wound infection in elective colorectal surgery by local application of a gentamicin-containing collagen sponge. Eur. J. Sur. Suppl. Acta Chir., 31–35 (1997)
86. L.G. Jorgensen, T.S. Sorensen, J.E. Lorentzen, Clinical and pharmacokinetic evaluation of gentamycin containing collagen in groin wound infections after vascular reconstruction. Eur. J. Vasc. Surg. **5**, 87–91 (1991)
87. S. Govender, C. Csimma, H.K. Genant, A. Valentin-Opran, Y. Amit, R. Arbel, H. Aro, D. Atar, M. Bishay, M.G. Borner, P. Chiron, P. Choong, J. Cinats, B. Courtenay, R. Feibel, B. Geulette, C. Gravel, N. Haas, M. Raschke, E. Hammacher, D. van der Velde, P. Hardy, M. Holt, C. Josten, R.L. Ketterl, B. Lindeque, G. Lob, H. Mathevon, G. McCoy, D. Marsh, R. Miller, E. Munting, S. Oevre, L. Nordsletten, A. Patel, A. Pohl, W. Rennie, P. Reynders, P.M. Rommens, J. Rondia, W.C. Rossouw, P.J. Daneel, S. Ruff, A. Ruter, S. Santavirta, T.A. Schildhauer, C. Gekle, R. Schnettler, D. Segal, H. Seiler, R.B. Snowdowne, J. Stapert, G. Taglang, R. Verdonk, L. Vogels, A. Weckbach, A. Wentzensen, T. Wisniewski, Recombinant human bone morphogenetic protein-2 for treatment of open tibial fractures: A prospective, controlled, randomized study of four hundred and fifty patients J. Bone Joint Sur. American volume, 84-A, 2123–2134 (2002).

88. F. Lefebvre, P. Pilet, N. Bonzon, G. Daculsi, M. Rabaud, New preparation and microstructure of the EndoPatch elastin-collagen containing glycosaminoglycans. Biomaterials **17**, 1813–1818 (1996)
89. C.J. Doillon, F.H. Silver, Collagen-based wound dressing: effects of hyaluronic acid and fibronectin on wound healing. Biomaterials **7**, 3–8 (1986)
90. M. Chvapil, Considerations on manufacturing principles of a synthetic burn dressing: A review. J. Biomed. Mater. Res. **16**, 245–263 (1982)
91. I.V. Yannas, J.F. Burke, P.L. Gordon, C. Huang, R.H. Rubenstein, Design of an artificial skin. II. Control of chemical composition. J. Biomed. Mater. Res. **14**, 107–132 (1980)
92. Z. Wachol-Drewek, M. Pfeiffer, E. Scholl, Comparative investigation of drug delivery of collagen implants saturated in antibiotic solutions and a sponge containing gentamicin. Biomaterials **17**, 1733–1738 (1996)
93. B.A. Weissman, D.A. Lee, Oxygen transmissibility, thickness, and water content of three types of collagen shields. Arch. Ophthalmol. **106**, 1706–1708 (1988)
94. W. Friess, G. Lee, M.J. Groves, Insoluble collagen matrices for prolonged delivery of proteins. Pharm. Dev. Technol. **1**, 185–193 (1996)
95. C. Chai, K.W. Leong, Biomaterials approach to expand and direct differentiation of stem cells. Mol. Ther. J. Am. Soc. Gene Ther. **15**, 467–480 (2007)
96. S. Hong, H.J. Hsu, R. Kaunas, J. Kameoka, Collagen microsphere production on a chip. Lab Chip **12**, 3277–3280 (2012)

Chapter 3
Creating Tunable Collagen Matrices – An Approach Inspired by *In Vivo* Collagen Synthesis and Self-Assembly

3.1 Tuning Molecular Delivery Through Collagen

Many complex biological tissues in the human body, including collagen-based tissues, display some remarkable features in common, including molecular self-assembly, a hierarchical organization at the atomistic, molecular, and macroscales, as well as multifunctionality [1]. In the context of soft tissue engineering, multifunctionality means collagen platform should be capable of offering structural support to the new regenerating tissue, conferring mechanical strength to cells to proliferate, migrate, and function, and provide the desired rate of mass transport of a drug molecule. The tunability of collagen then refers to the ability to regulate such characteristics of collagen per need of a tissue, through careful control of collagen microstructure and proteolytic degradability, the two features that can affect fluid or solute transport through collagen. This kind of tunability provides us with an opportunity to guide cell fate and tissue formation in the body, as illustrated in Fig. 3.1. Thus, the salient features of a multifunctional drug delivery platform should be its customizability, ability to induce cellular signaling and tissue integration, ability to show a broad range of mechanical properties, and format versatility, all possibly in the absence of exogenous-crosslinking.

Though collagen processing and the resulting materials had been investigated for decades, engineering production variants have not all been explored [2] and applied toward drug delivery. One reason is that much of the collagen engineering technology is based on traditional knowledge, but another reason is that the collagen structure and material's behavior depends on as many surrounding conditions as humidity, kinds, and concentrations of buffers, and the stability of the structure of all levels including crosslinking. Even low amounts of additives and/or processing of collagen with chemical/enzymatic cross-links influence the structural, chemical, physical, and biological materials properties of collagen [2].

© The Author(s), under exclusive license to Springer Nature Switzerland AG 2021
R. Joshi, *Collagen Biografts for Tunable Drug Delivery*, SpringerBriefs in
Applied Sciences and Technology, https://doi.org/10.1007/978-3-030-63817-7_3

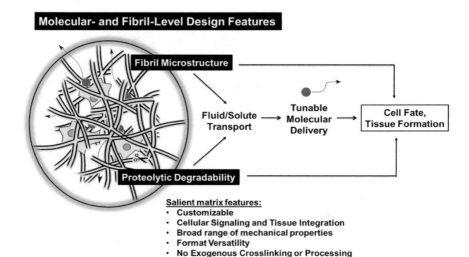

Fig. 3.1 Schematic of collagen-based drug delivery platform with tunable fibrillar microstructure and proteolytic degradability. These tunable features affect (1) solute/fluid transport and (2) cell fate and tissue formation. The orange spheres represent drug molecules released from collagen fibril-matrix

Recent years also saw a rise in the decellularization of whole organs and organ parts as a routine technique to prepare complex structures of the ECM by saving the original structure including the vessel patterns. While this complex process preserves the collagen structure in ECM, it is challenging to recellularize such architectures, not to mention the expected requirements of purity and quality control if such materials are aimed to be accepted by the authorities and used by surgeons [2]. Therefore, the field between complex decellularized organs and the marketed simple shapes such as films, sponges, and powders of collagen is open for new engineering techniques based on purified intermediates.

From a drug delivery perspective, to control molecular transport through collagen-based matrices in vitro, it is important to mimic collagen self-assembly and provide control over the microstructure and proteolytic degradability of resultant matrices. However, the conventional monomeric collagen (telocollagen and atelocollagen) formulations fail to capture the self-assembly characteristic of collagen as we saw in Chap. 2. Unlike oligomers, these conventional monomeric formulations do not retain their tissue-specific, covalent intermolecular cross-links [3]. As a result, the matrices formed from conventional collagen formulations display weaker mechanical integrity and rapid proteolytic degradation. Consequently, they fail to retain molecules/drugs for a longer time and cannot be tuned to match their release rates to the desired rate.

A lot has been written about modifying collagen structures, strength, and stability by exogenous cross-linking [4–10]. However, due to the disadvantages of altering the collagen native structure that these methods bring, here we look at alternative engineering variables. Instead of relying on chemically cross-linking traditional collagen formulations, molecular release from collagen-based materials can be tuned through a systematic variation of *collagen polymer composition* and *fibril density*. The basis of varying the molecular release from collagen is a modulation of *microstructure* and *proteolytic degradability* features of the collagen matrix, as these two happen to be the main regulators of molecular transport.

3.2 Tuning Microstructure and Molecular Release Properties of Oligomer Fibril Matrices

Collagen implants should be tunable in terms of their mechanical, microstructural, and proteolytic degradation properties to match tissue regeneration rates, as these rates can vary based on the location of soft tissue damage or its healing stage, as well differences in an individual's age, dietary intake, healing rate, and lifestyle-related factors [11]. Despite wide research and promising results of collagen-based materials in improving therapeutic efficacy and delivery [10, 12, 13], the inability of the collagen-based systems to provide tunable release is still a major limitation restricting its clinical utility.

Challenges in tuning molecular release from conventional collagen matrices stem from the open weave structure of collagen [10]. The conventional formulations have a poor molecular composition and inability to fully capitalize on the inherent self-assembly of collagen leading to weak mechanical integrity and cursory control over physical and molecular release properties [14–17]. Hence, to tune molecular release from collagen, many scientists rely on methods such as exogenous cross-linking, mixing with another polymer phase, covalent or non-covalent bonding, or sequestration in a secondary matrix, etc. The state-of-the-art strategies in making controlled release systems from collagen are listed in Appendix I – Table A.

While such strategies have resulted in limited success, the steps involved not only increase the complexity of the system but also affect the microstructure of collagen [18, 19]. Therefore, a combination of improved collagen formulation and a tuning strategy that does not alter physiologically relevant properties of collagen is desirable to improve tunable molecular delivery from collagen. One such strategy is to use unique self-assembling collagen building blocks such as oligomers, telocollagens, and atelocollagens in a combination to tune release. For example, a self-assembling collagen-based drug delivery system consisting of different amounts of atelocollagen and oligomers can be formulated [20], and the specific molecular release properties of the resulting can be tuned by tweaking the amount of oligomer or atelocollagen. The microstructures of telocollagen and oligomer matrices are different and display different stiffnesses as evidenced by Bailey et al. [16] by the difference in the shear storage and compressive moduli of oligomer and telocollagen matrices. The differences in tunability and the 3D structure of matrices made from telocollagen (monomer) and oligomer as reported by Puls et al. [21] is indicated in Table 3.1. Such differences can be exploited to mix the various collagen building blocks in different proportions to affect the molecular release.

Table 3.1 Differences in the characteristics of telocollagen and oligomer

Characteristics	Monomeric (Telocollagenic) Type I collagen matrix	Oligomeric Type I collagen matrix
Source	Rat tail or bovine tendon	Pig dermis
Primary component(s)	Type I collagen monomers	Type I collagen oligomers
3D structure	Entanglement of long fibrils with little to no mature intermolecular crosslinks	Highly interconnected fibril matrix with mature intermolecular crosslinks
User tunabilty	Moderate	Good
Mechanical stability	Poor	Good
Range of achievable stiffness	1–4 mg/ml: 9–343 Pa	1–4 mg/ml: 27–1440 Pa

Modified from [24]

Polymerization time can also be affected by the combination of various building blocks of collagen in different ratios. A shorter polymerization time would be beneficial if collagen-based drug delivery device is needed to be injected and then polymerized at the site of application. Joshi [22] investigated the effect of varying ratio of oligomer and atelocollagen building blocks upon their polymerization. They combined oligomer and atelocollagen building blocks in the ratio of 0:100%, 25:75%, 50:50%, 75:25%, and 100:0% and looked at the time of polymerization of resultant matrices. They observed that the stiffness of matrices increased with increased percentage of oligomer in oligomer: atelocollagen combinational ratios (Fig. 3.2a). The rate of polymerization increased (Fig. 3.2b), and T_{half} of polymerization (Time to 50% of polymerization) decreased (Fig. 3.2c) as oligomer% went up in the oligomer: atelocollagen ratio.

Interestingly, all the matrix combinations polymerized with a T_{half} of polymerization < 5 min, except for the 100% atelocollagenic matrix (oligomer: atelocollagen ratio = 0:100) (Fig. 3.2c). The rapid polymerization is an important design feature in clinical applications, where it might be necessary that injected collagen solutions polymerize quickly to create a solid matrix. This may allow for appropriate matrix placement and molecular delivery in situ, a feature not always exhibited by many conventional collagen formulations made up of atelocollagen [14].

3.3   Tunability of Molecular Release Through Collagen Densification

Increasing fibril density or collagen concentration is another way to control molecular release as this approach decreases matrix porosity [10, 23]. For example, microstructurally, the increase of oligomer concentration from 0.9 to 2.1 mg/ml was observed to reduce the void space while enhancing the collagen matrix stiffness, as illustrated by Puls et al. [24] in Fig. 3.3.

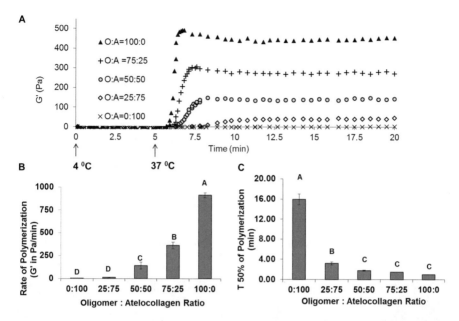

Fig. 3.2 Collagen polymerization kinetics is dependent upon the oligomer: atelocollagen ratio. Time-dependent changes in shear-storage modulus were monitored as collagen formulations exhibited solution to matrix transition following an increase in temperature from 4 °C to 37 °C. The polymerization profiles (**a**) were used to quantify the initial rate of polymerization (mean ± SD, **b**) and half-polymerization times (mean ± SD, **c**) for N = 3 of each matrix type

This strategy has been previously applied to control the release of several molecules from collagen-based materials. For example, by varying collagen content from 1.5% to 2.0% and 2.5%, FITC-coupled pexiganan release from collagen was extended from 24 h to 48 h and 72 h, respectively [25]. Lauzon et al. observed that modulating the concentration of collagen hydrogels from 1.5 to 4.5 mg/ml affected pBMP-9 interaction with collagen and its molecular release [26]. Fujioka et al. [27–29] observed a sustained release of various proteins by increasing collagen density using methods such as ethanol immersion and air drying. While these methods were successful in tuning molecular release, an associated concern is a decrease in the porosity to an extent that cell migration and proliferation can be hampered [10]. Therefore, it is important to maintain a balance between matrix porosity and collagen concentration.

Another problem associated with densifying collagen matrices is the viscosity of collagen. Practically, due to the high viscosity, formulating collagen solutions at concentrations above 10% has been very difficult [28, 30, 31]. Therefore, alternative methods such as reverse dialysis [32], continuous injection, and evaporation [33] and centrifugation followed by polymerization [34] have been attempted to increase the density of collagen. Unfortunately, these methods can require weeks to months to prepare and can result in matrices with varying microstructures [17], and limited cell migration or infiltration into the densified material [35, 36].

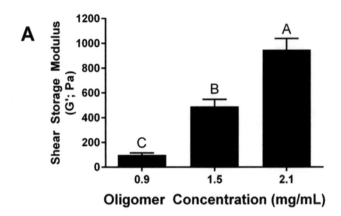

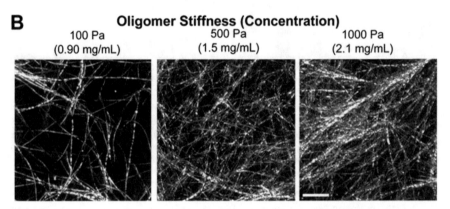

Fig. 3.3 (**a**) Matrix stiffness values of matrices prepared at Oligomer concentrations of 0.9, 1.5, 2.1 mg/mL are given as shear storage modulus (G'; mean ± SD) with letters indicating statistically different groups (p < 0.05, n = 3). (**b**) Images represent z-stack projections of confocal reflection microscopy (10 μm thickness, scale bar = 10 μm) of matrices. (Figure adapted from work by T.J. Puls et al. [24], https://doi.org/10.1371/journal.pone.0188870, covered under the creative commons license https://creativecommons.org/licenses/by/4.0/)

To overcome these limitations, and to better approximate the structural hierarchy and mechanical properties of mature tissues, Blum et al. recently used the method of confined compression on oligomeric collagen matrices to yield high-fibril density matrices with high cell viability [17]. Due to the success of this technique in maintaining collagen microstructure and physiological relevance (D banding pattern) even at high fibril density, this strategy seems attractive to apply it for molecular release. Increasing collagen concentration has been correlated with an increase in fibril density and a concomitant increase in matrix stiffness [14, 16, 17, 37, 38]. Such an increase in fibril density is seen to affect collagen microstructure as indicated by the cryo-SEM images of oligomeric collagen without confined compression (Fig. 3.4a) and with confined compression (Fig. 3.4b, c). Since increased fibril density is known to affect molecular transport from collagen, the molecular release

Scale bar = 10 μm

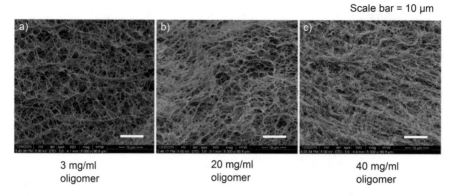

3 mg/ml
oligomer

20 mg/ml
oligomer

40 mg/ml
oligomer

Fig. 3.4 Fibril microstructure of (**a**) 3 mg/ml, (**b**) 20 mg/ml, and (**c**) 40 mg/ml oligomer matrices as visualized using cryo-SEM. Scale bar represents 10 μm. (Own results, unpublished)

may be affected through the use of collagen matrices densified via confined compression, as indicated in Fig. 3.5, based on Joshi's work [22]. The data in Fig. 3.5 shows that FITC-Dextran release observed in compressed matrices was significantly more sustained compared to non-compressed matrices for both 10 KDa (A-C) and 2 MDa (D-F) FITC-Dextrans.

3.4   Tunability of Molecular Release Through Modulation of Proteolytic Degradability

For clinical applications, collagen-based biomaterials must overcome any fragility; they should be easy to handle and be applicable by surgical techniques, and their degradability and tissue integration have to follow the regeneration of the surrounding tissue. The key properties in ensuring this are mechanical stability and enzymatic degradability of collagen, which in turn are strongly influenced by mechanical disintegration, and denaturation of the triple helix and cross-linking.

The collagen triple helix is unsusceptible to many proteases except collagenases. Decellularized porcine dermis acts as an effective substrate for bacterial collagenase, and with decreasing susceptibility for Pronase, Termolysin, and Proteinase K. Once the triple helices of collagen are denatured, the protein chains are easily susceptible to the degradation by various other enzymes such as Pronase, Termolysin, and Proteinase within a very short period [2]. In our bodies, enzymes responsible for the collagen degradation in the extracellular matrix (ECM) are matrix metalloproteinases (MMPs). At the fibrillar level, degradation results in decreased mechanical strength of matrices, and rapid release of molecules if drug-loaded. Furthermore, the degradation of collagen as one of the components of the ECM is a very important process in the development, morphogenesis, tissue remodeling, and repair. It is tightly regulated in physiological conditions, and its dysregulation triggers many

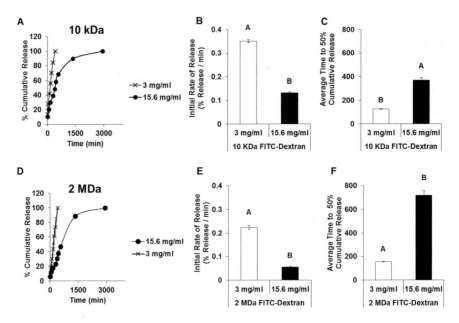

Fig. 3.5 Densifying collagen fibril matrices modulates molecular release profiles. 10 kDa and 2 MDa FITC-Dextran containing 3 mg/ml oligomeric matrices were compressed and their release characteristics were measured upon exposure to 50 U/ml collagenase. The difference between release profiles given by low-density and high-density matrix is enhanced for 2 MDa FITC-Dextran (**d**) as compared to 10 kDa FITC-Dextran (**a**). The initial rate of release is defined as the slope of the graphs before the cumulative release profiles reached 50% release in (**a**) and (**d**). The initial rate of release is significantly lower in high-density matrices compared to low-density matrices for both 10 KDa (**b**) and 2 MDa (**e**) FITC-Dextrans. The T50% of release (Time to 50% of molecular release) is significantly higher in high-density matrices, than in low-density matrices for both smaller size FITC-Dextran (10 kDa) and larger size (2 MDa) (**c** and **f**). (Figure based on unpublished own-work)

diseases such as cancer, rheumatoid arthritis, nephritis, encephalomyelitis, chronic ulcers, and fibrosis [39]. Therefore, when designing collagen materials for controlled release, special attention is needed on tunability of the proteolytic degradability of collagen.

The tunability of molecular release from collagen must be tested under proteolytic conditions. Proteolytic degradation of collagen is seen in pathological conditions such as cancer [40, 41], chronic foot ulcers [42] or, even during normal tissue homeostasis, and is caused by the members of matrix metalloproteinase (MMP) family. One can test drug release in vitro from a newly designed collagen-based material by exposing drug-loaded materials to buffer solutions such as Phosphate Buffer Saline (PBS), however, doing so in absence of collagenase only reveals a diffusion-based release characteristic. It does not mimic in vivo molecular release from collagen that occurs through a combination of diffusion and enzymatic breakdown of the collagen matrix [43]. To test a collagen-based material's true molecular release characteristics, one should therefore validate the molecular release

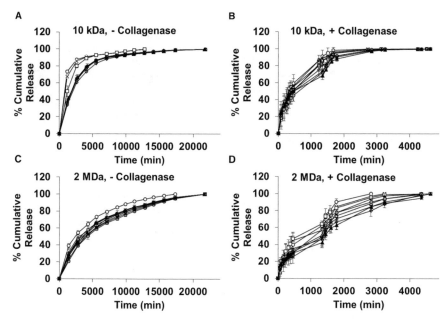

Fig. 3.6 Variation in molecular release profiles obtained by mixing oligomer and atelocollagen in different ratios is enhanced in the presence of collagenase. FITC-dextrans (10 kDa and 2000 kDa) were admixed within oligomer: atelocollagen solutions prepared at ○ (0:100), × (5:95), □ (10:90), Δ (15:85), + (20:80), ◇ (25:75), ● (50, 50), * (75,25), and ▲ (100,0) ratios and, upon polymerization, time-dependent release of FITC-dextrans was monitored spectrofluorometrically in the absence (**a** & **c**) and presence (**b** & **d**) of 10 U/ml collagenase. (Unpublished own-work)

by exposing collagen to an appropriate amount of collagenase buffer that can induce diffusion-based release as well as mimic the in vivo matrix metalloproteinase (MMP)-based degradation mechanism. It should be noted that the tunability of the molecular release obtained in absence of collagenase can be different from the tunability obtained in presence of collagenase, as indicated by Fig. 3.6.

Even without exposure to collagenase, many collagen-based drug-delivery devices show rapid degradation [10] and therefore, unwanted rapid release of drug molecules encapsulated. Conventional collagen formulations generally display a rapid release of drug molecules owing to their rapid proteolytic degradation. For example, basic fibroblast growth factor (bFGF) was released from non-cross-linked collagen matrices during the first 6 h [44]. Similarly, a collagen sponge incubated with rhBMP-2 (~26 kDa) solution released 55% of the protein in 1 h and 100% in 2 days [45]. Larger molecule Riboflavin (376.36 g/mol) was also released in a short duration of 16 h from collagen sponges [46]. Implantation of a gentamicin-impregnated collagen sponge Garamycin in horses resulted in the peak concentration of gentamicin within 3 h [47]. These rapid molecular release examples from conventional collagen indicate the necessity to design collagen materials that can overcome the limitations of conventional formulations and provide tunable and sustained molecular release.

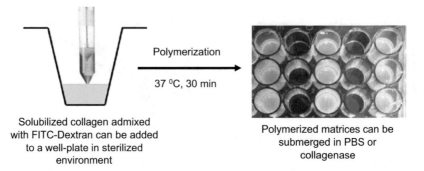

Solubilized collagen admixed with FITC-Dextran can be added to a well-plate in sterilized environment

Polymerization

37 °C, 30 min

Polymerized matrices can be submerged in PBS or collagenase

Fig. 3.7 In vitro set up for measurement of release kinetics from collagen. Solubilized collagen with FITC-Dextran in it can be inserted using a positive displacement micropipette in a well-plate and allowed to polymerize at 37 °C. After polymerization, collagen matrices formed in the well plate can be exposed to buffer such as PBS with or without collagenase and drug release be monitored by periodic replacement of buffer solution on the top of matrices

To overcome rapid proteolytic degradation and control drug release kinetics, scientists have applied various strategies to control the proteolytic degradability of collagen matrices, such as modification of the matrix characteristics (porosity, density), and use of different chemical treatment regimes that cross-link collagen matrices for stability [10, 23]. Another interesting strategy is loading collagen matrices with MMP inhibitors. For example, collagen matrices were impregnated with Siderophore-loaded Gelatin Microspheres (SGM) as a delivery system to control both infection and protease levels in the wound site for accelerated healing [48]. Siderophore is an ion chelating agent used to inhibit MMPs at the wound sites. Proteolytic degradation of collagen is thus an important parameter to consider in designing tunable matrices.

While many studies have documented controlled degradation of collagen-based materials [49, 50] and drug delivery devices [51–53], these were conducted at only one level of collagenase. Literature about collagen-based drug delivery devices providing controlled degradation at varying collagenase levels is sparse [51, 54]. Moreover, data for enzymatic activity is usually limited. An in vitro experiment should test molecular release from a collagen-based material at various collagenase levels by keeping the actual target site of implantation and its pathological characteristics in mind. This is because studies have indicated that the collagenase levels vary in normal versus pathophysiological states and also at various locations *in vivo* [42, 54]. Moreover, the level of collagenase varies according to the age of the wound [55]. Therefore, collagen-based implants must be tunable in terms of their molecular release under varying collagenase levels [56, 57].

To capture collagenase-based molecular release, successful in vitro experimental models have been established and used in the past by several researchers for quantifying molecular release from various collagen-based drug delivery systems

[58–65]. In such models, the drug-containing collagen matrices are typically submerged in a small volume of PBS buffer (typically 400–2500 µL) containing collagenase at various levels, and the system is subjected to gentle shaking. The buffer volume is replaced periodically in the given release study period, to quantify drug elution at various time points. An illustration of such a strategy based on work by Joshi [20] is shown in Fig. 3.7.

3.5 Empirical Modeling of Drug Release

Empirical modeling of drug release plays an important role in understanding the mechanisms of drug release [66–68]. Characterizing molecular release from polymer matrices has been accomplished through the use of various empirical models, including the well-known Higuchi and Peppas and Weibull models [69–76]. The basic mathematical expressions used to describe the release kinetics and discern the release mechanisms are elegantly described in the articles on Higuchi law [77] and the Peppas equation or the so-called power law ([78–80]), as well as the recent review on mathematical modeling of release kinetics from supramolecular drug delivery systems [81]. It is noted form literature review that despite a wide use [74–76, 82, 83], both Higuchi and power-law are short-time approximations of complex exact relationships [83, 84], therefore, their use is confined to the description of the first 60% of the release curve [82]. Beyond 60%, the quality of the fit has been observed to be poor. However, the Weibull function can be appropriate for fitting almost the entire set of data while effectively explaining the mechanisms of molecular release [82, 85–88]. Weibull model has been applied successfully by several researchers for the discernment of drug release mechanisms [86–98]. Furthermore, Papadopoulou [82] described how the different Weibull parameters can be correlated with various release mechanisms, as illustrated in Table 3.2. Through the study of experimental data, Papadopoulou et al. [82] concluded that ß

Table 3.2 Exponent b of Weibull function and mechanism of release described by Papadopoulou [82]

b	Release mechanism remarks
b < 0.35	May occur in highly disordered spaces much different than the percolation cluster
b ~ 0.35–0.39	Diffusion in fractal substrate morphologically similar to the percolation cluster
0.39 < b < 0.69	Diffusion in fractal or disordered substrate different from the percolation cluster
b ~ 0.69–0.75	Diffusion in normal Euclidian space
0.75 < b < 1	Diffusion in normal Euclidian substrate with contribution of another release mechanism
b = 1	First-order release obeying Fick's first law of diffusion
b > 1	Sigmoid curve indicative of complex release mechanism

is an indicator of the mechanism of transport of the drug through the polymer matrix; $\beta \leq 0.75$ indicates Fickian diffusion, while value sin the range of $0.75 < \beta < 1$ are associated with a combined mechanism (Fickian diffusion and swelling-controlled transport). For values of β higher than 1, the drug transport follows a complex release mechanism.

In closing, we realize that mathematical models will continue to help us predict the properties of processed collagen materials for drug delivery applications before the material can be applied clinically. The therapeutic use of collagen as a drug delivery device can be improved by regulating the engineering variables of micro-structure and proteolytic degradability. The next challenge is then to engineer highly complex, vascularized structures of tissue using a collagen matrix while promoting the native biological systems to reorganize themselves. This means that cells of the recipient *in vivo* are capable of building functional capillaries, veins, or nerves. The biological graft systems developed using collagen must have the potential to assemble simple molecules into complicated architectures on the microscopic level which follow the complex physiological demands and the requirements of the individual. The use of collagen as material for manufacturing drug delivery devices is only at the beginning. As discussed in this chapter, the combinations of different collagen-building blocks, confined compression methods as well as collagen-based peptides and the combination with other ECM-derived polymers may promise challenging developments [2].

In the next chapter, we will talk about the collagen-based materials in a specific therapeutic application – wound-healing. Complex and chronic wounds, such as skin ulcers, are difficult to heal, and pose a major clinical challenge for our society, increasingly impacting the health and lifestyle of those suffering and their families. At present, 6.5 million people are affected by chronic wounds in the United States alone [99] with an estimated 25 billion dollars spent annually to treat these patients. This societal and economic burden can be reduced by an intervention from an alternative multifunctional tissue engineering strategy based on collagen, a molecule native to our own body.

Appendix I

Table A State-of-the-art methods of tuning collagen based molecular release

Strategy	Example	Molecule delivered	Release period	References	Limitation
Varying extent of exogenous crosslinking	Crosslinking with glutaraldehyde	Vascular endothelial growth factor (VEGF)	30 days	[100]	Detrimental effects on cells and tissues, such as cytotoxicity or tissue calcification
	Crosslinking with four-arm poly (ethylene glycol)-terminated succinimidyl glutarate (4S-StarPEG)	siRNA	10 days	[101]	Release requires hydrolysis of a linking bond which is different from *in vivo* proteolytic degradation
	Crosslinking with N-(3-dimethylaminopropyl)-N'-ethylcarbodiimide (EDC) and N-hydroxysuccinimide (NHS)	Heparin	11 days	[102]	Crosslinking with additives increases complexity of system
	Crosslinking with metal oxide nanoparticles (NPs) and PVP capped ZnO (ZnO/PVP) in addition to UV crosslinking	Pilocarpine hydrochloride (PHCl)	14 days	[103]	Crosslinking reagents can also react with and affect non-collagen structural proteins, glycosaminoglycans, growth factors and other bioactive compounds, or cells
Chemical modification of collagen to enable ionic bonding between drug and collagen	Succinylating collagen sponge and film with drug dispersed in poly (N-vinyl-2-pyrrolidione) (PVP) solution	Ciprofloxacin (a cationic fluoroquinilone antibiotic)	5 days	[104]	Succinylated collagen gels do not appear to have a long lifetime *in vivo*, usually disappearing within 24 h depending on the degree of succinylation

(continued)

Table A (continued)

Strategy	Example	Molecule delivered	Release period	References	Limitation
Covalent immobilization of drug	Crosslinking with difunctional or multifunctional succinimidyl ester polyethylene glycol (PEG, 3.4 to 10 kDa)	Transforming growth factor beta-2 (TGF-beta2)	5 days	[105]	Covalent conjugation can be difficult to control, produce poor reaction yields, and even compromise the biochemical features of the protein/drug or collagen itself
	Crosslinking with SS-PEG-SS	VEGF	72 h	[106]	
	Poly(dialdehyde) guar gum (PDAGG)-based covalent crosslinking of biomolecules with collagen	Platelet-derived growth factor (PDGF)	13 days	[107]	
Adding intermediate proteins with affinity for collagen and protein of interest	Heparin	Basic fibroblast growth factor (bFGF)	10 days	[108]	Binding interactions are specific to each drug and hard to predict
	Fibronectin	Recombinant human bone morphogenic factor 2 (rhBMP-2)	7 days	[109]	Very little tuning if the binding interaction is weak

Mixing collagen with other synthetic or natural polymers	Hybrid scaffolds of collagen and poly (lactic-co-glycolic acid) microbeads were prepared by introducing insulin-releasing poly (lactic-co-glycolic acid) microbeads into collagen porous scaffolds. Pore structure was controlled using ice particulates	Insulin	4 weeks	[110]	Reduction in material's cell-instructive capacity and its inability to integrate with host tissue can make the clinical translation of products very difficult
	Collagen–hydroxyapatite scaffolds combined with either alginate or poly(lactic-co-glycolic acid) (PLGA) microparticles	rhBMP-2	28 days	[111]	
	Addition of BMP-2 into soft PEG hydrogels before infusion in to the solid collagen/HA sponges	BMP-2	40% release observed in 15 days	[112]	
	Collagen and poly-(caprolactone)	Gentamicin and amikacin	60 h	[113]	
	Collagen impregnated with drug loaded alginate microspheres	Antibacterial agent silver sulfadiazine (AgSD)	66.8% released in 72 h	[114]	
	Lyophilizing solution of suspended PLGA microparticles in a collagen dispersion	Gentamicin	7 days	[115]	
	Drug containing liposome sequestration in collagen gel	Insulin	5 day	[116]	
		Growth hormone	14 day		
Engineering peptides with collagen binding domain (CBD)	Fusion protein consisting of hepatocyte growth factor (HGF; an angiogenic factor) and a collagen-binding domain (CBD) polypeptide of fibronectin was produced in a baculovirus expression system	Hepatocyte growth factor (HGF)	1 week	[117–124] [117]	Complexity of such systems is a disadvantage from a commercial perspective

(continued)

Table A (continued)

Strategy	Example	Molecule delivered	Release period	References	Limitation
Addition of collagen mimetic peptides (CMPs)	CMP-modified polyplexes are bound to collagen via thermally induced annealing that induces CMP strand invasion and CMP-collagen triple helical hybridization	Gene	1 month	[125]	Native collagen microstructure is modified
Vitrification of collagen membrane	Collagen gel was dried for 2 weeks to convert into a rigid glass-like material, which was rehydrated with PBS containing VEGF	VEGF	14 days	[126]	Native collagen microstructure is modified
	After gelation, collagen membranes were formed by vitrification for 2 days, followed by rehydration with PBS containing BMP-2	BMP-2	> 80% retained even after 15 days	[127]	
Increasing collagen density	Collagen content varied from 1.5% to 2.0% and 2.5%, pexiganan release from collagen was found to be extended from 24 h to 48 h and 72 h, respectively	FITC-coupled Pexiganan (a 22 amino acid antimicrobial peptide)	72 h	[128]	Can limit cell migration/ infiltration into the densified collagen
	A membrane consisting of photo-polymerized polyethylene glycol dimethacrylate (PEGDM) and interconnected collagen microparticles (COLs) was used and collagen concentration varied from 100 mg/ml to 300 and 500 mg/ml	40-kDa FITC–dextran, and recombinant human brain-derived neurotrophic factor (rhBDNF)	42 days	[129]	
	Concentration of type I collagen hydrogels was varied from 1.5 to 4.5 mg/ml; drug interaction also played a role in tuning the release	pBMP-9	72 h	[130]	

References

1. P.A. Vasquez, M.G. Forest, Complex fluids and soft structures in the human body, in *Complex fluids in biological systems: Experiment, theory, and computation*, ed. by E. S. Spagnolie, (Springer, New York, 2015), pp. 53–110
2. M. Meyer, Processing of collagen based biomaterials and the resulting materials properties. Biomed. Eng. Online **18**, 24–24 (2019)
3. S.L. Voytik-Harbin, B. Han, Collagen-cell interactions in three-dimensional microenvironments, in *Handbook of imaging in biological mechanics*, (CRC Press, 2014), pp. 261–274
4. Q. Jiang, N. Reddy, S. Zhang, N. Roscioli, Y. Yang, Water-stable electrospun collagen fibers from a non-toxic solvent and crosslinking system. J. Biomed. Mater. Res. A (2012)
5. D.T. Cheung, M.E. Nimni, Mechanism of crosslinking of proteins by Glutaraldehyde II. Reaction with monomeric and polymeric collagen. Connect. Tissue Res. **10**, 201–216 (1982)
6. J.M. Lee, H.H.L. Edwards, C.A. Pereira, S.I. Samii, Crosslinking of tissue-derived biomaterials in 1-ethyl-3-(3-dimethylaminopropyl)-carbodiimide (EDC). J. Mater. Sci. Mater. Med. **7**, 531–541 (1996)
7. J.M. Lee, C.A. Pereira, L.W.K. Kan, Effect of molecular-structure of Poly(Glycidyl Ether) reagents on cross-linking and mechanical-properties of bovine pericardial xenograft materials. J. Biomed. Mater. Res. **28**, 981–992 (1994)
8. R. Tu, S.H. Shen, D. Lin, C. Hata, K. Thyagarajan, Y. Noishiki, R.C. Quijano, Fixation of bioprosthetic tissues with monofunctional and multifunctional polyepoxy compounds. J. Biomed. Mater. Res. **28**, 677–684 (1994)
9. L.H.H. Olde Damink, P.J. Dijkstra, M.J.A. van Luyn, P.B. van Wachem, P. Nieuwenhuis, J. Feijen, Cross-linking of dermal sheep collagen using a water-soluble carbodiimide. Biomaterials **17**, 765–773 (1996)
10. W. Friess, Collagen–biomaterial for drug delivery, Eur. J. Pharm. Biopharm. **45**, 113–136 (1998)
11. Q.L. Loh, C. Choong, Three-dimensional scaffolds for tissue engineering applications: Role of porosity and pore size. Tissue Eng. Part B, Rev. **19**, 485–502 (2013)
12. C.H. Lee, A. Singla, Y. Lee, Biomedical applications of collagen. Int. J. Pharm. **221**, 1–22 (2001)
13. D.G. Wallace, J. Rosenblatt, Collagen gel systems for sustained delivery and tissue engineering. Adv. Drug Deliv. Rev. **55**, 1631–1649 (2003)
14. S.T. Kreger, B.J. Bell, J. Bailey, E. Stites, J. Kuske, B. Waisner, S.L. Voytik-Harbin, Polymerization and matrix physical properties as important design considerations for soluble collagen formulations. Biopolymers **93**, 690–707 (2010)
15. P.J. Critser, S.T. Kreger, S.L. Voytik-Harbin, M.C. Yoder, Collagen matrix physical properties modulate endothelial colony forming cell-derived vessels in vivo. Microvasc. Res. **80**, 23–30 (2010)
16. J.L. Bailey, P.J. Critser, C. Whittington, J.L. Kuske, M.C. Yoder, S.L. Voytik-Harbin, Collagen oligomers modulate physical and biological properties of three-dimensional self-assembled matrices. Biopolymers **95**, 77–93 (2011)
17. K.M. Blum, T. Novak, L. Watkins, C.P. Neu, J.M. Wallace, Z.R. Bart, S.L. Voytik-Harbin, Acellular and cellular high-density, collagen-fibril constructs with suprafibrillar organization. Biomater. Sci **4**, 711–723 (2016)
18. S. Lin, L. Hapach, C. Reinhart-King, L. Gu, Towards tuning the mechanical properties of three-dimensional collagen scaffolds using a coupled fiber-matrix model. Materials **8**, 5254 (2015)
19. M.A. Urello, K.L. Kiick, M.O. Sullivan, A CMP-based method for tunable, cell-mediated gene delivery from collagen scaffolds. J. Mater. Chem. B **2**, 8174–8185 (2014)
20. R. Joshi, Designer collagen-fibril biograft materials for tunable molecular delivery (2016)

21. T.J. Puls, X. Tan, M. Husain, C.F. Whittington, M.L. Fishel, S.L. Voytik-Harbin, Development of a novel 3D tumor-tissue invasion model for high-throughput. High-Content Phenotypic Drug Screen., Sci. Rep. **8**, 13039–13039 (2018)
22. R. Joshi, Designer collagen-fibril biograft materials for tunable molecular delivery, Open Access Dissertations, in: Biomedical Engineering, Purddue University, West Lafayette, Inddiana (2016)
23. Z. Ruszczak, W. Friess, Collagen as a carrier for on-site delivery of antibacterial drugs. Adv. Drug Deliv. Rev. **55**, 1679–1698 (2003)
24. T. Puls, X. Tan, C.F. Whittington, S.L. Voytik-Harbin, 3D collagen fibrillar microstructure guides pancreatic cancer cell phenotype and serves as a critical design parameter for phenotypic models of EMT. PLoS One **12**, e0188870 (2017)
25. D. Gopinath, M.S. Kumar, D. Selvaraj, R. Jayakumar, Pexiganan-incorporated collagen matrices for infected wound-healing processes in rat. J. Biomed. Mater. Res. A **73A**, 320–331 (2005)
26. M.-A. Lauzon, B. Marcos, N. Faucheux, Effect of initial pBMP-9 loading and collagen concentration on the kinetics of peptide release and a mathematical model of the delivery system. J. Control. Release **182**, 73–82 (2014)
27. K. Fujioka, Y. Takada, S. Sato, T. Miyata, Novel delivery system for proteins using collagen as a carrier material: The minipellet. J. Control. Release **33**, 307–315 (1995)
28. K. Fujioka, M. Maeda, T. Hojo, A. Sano, Protein release from collagen matrices. Adv. Drug Deliv. Rev. **31**, 247–266 (1998)
29. M. Maeda, K. Kadota, M. Kajihara, A. Sano, K. Fujioka, Sustained release of human growth hormone (hGH) from collagen film and evaluation of effect on wound healing in db/db mice. J. Control. Release **77**, 261–272 (2001)
30. F. Peter, Collagen structure and mechanics, in *Collagen Structure and Mechanics: An Introduction*, ed. by F. Peter, (Springer Science + Business Media, LLC, New York, 2008)
31. G.C. Wood, M.K. Keech, The formation of fibrils from collagen solutions. 1. The effect of experimental conditions: Kinetic and electron-microscope studies. Biochem. J. **75**, 588–598 (1960)
32. N. Saeidi, K.P. Karmelek, J.A. Paten, R. Zareian, E. DiMasi, J.W. Ruberti, Molecular crowding of collagen: A pathway to produce highly-organized collagenous structures. Biomaterials **33**, 7366–7374 (2012)
33. G. Mosser, A. Anglo, C. Helary, Y. Bouligand, M.-M. Giraud-Guille, Dense tissue-like collagen matrices formed in cell-free conditions. Matrix Biol. **25**, 3–13 (2006)
34. J.L. Puetzer, L.J. Bonassar, High density type I collagen gels for tissue engineering of whole menisci. Acta Biomater. **9**, 7787–7795 (2013)
35. K.R. Johnson, J.L. Leight, V.M. Weaver, Demystifying the effects of a three-dimensional microenvironment in tissue morphogenesis. Methods Cell Biol. **83**, 547–583 (2007)
36. A. Nyga, M. Loizidou, M. Emberton, U. Cheema, A novel tissue engineered three-dimensional in vitro colorectal cancer model. Acta Biomater. **9**, 7917–7926 (2013)
37. A.M. Pizzo, K. Kokini, L.C. Vaughn, B.Z. Waisner, S.L. Voytik-Harbin, Extracellular matrix (ECM) microstructural composition regulates local cell-ECM biomechanics and fundamental fibroblast behavior: A multidimensional perspective. J. Appl. Physiol. **98**, 1909–1921 (2005)
38. B.A. Roeder, K. Kokini, J.E. Sturgis, J.P. Robinson, S.L. Voytik-Harbin, Tensile mechanical properties of three-dimensional type I collagen extracellular matrices with varied microstructure. J. Biomech. Eng. **124**, 214–222 (2002)
39. A. Jabłońska-Trypuć, M. Matejczyk, S. Rosochacki, Matrix metalloproteinases (MMPs), the main extracellular matrix (ECM) enzymes in collagen degradation, as a target for anticancer drugs. J. Enzyme Inhib. Med. Chem. **31**, 177–183 (2016)
40. H. Nagase, J.F. Woessner, Matrix Metalloproteinases. J. Biol. Chem. **274**, 21491–21494 (1999)

41. L.M. Coussens, Z. Werb, Matrix metal loproteinases and the development of cancer. Chem. Biol. **3**, 895–904

42. R. Lobmann, A. Ambrosch, G. Schultz, K. Waldmann, S. Schiweck, H. Lehnert, Expression of matrix-metalloproteinases and their inhibitors in the wounds of diabetic and non-diabetic patients. Diabetologia **45**, 1011–1016 (2002)

43. F.A. Radu, M. Bause, P. Knabner, G.W. Lee, W.C. Friess, Modeling of drug release from collagen matrices. J. Pharm. Sci. **91**, 964–972 (2002)

44. M.J.B. Wissink, R. Beernink, A.A. Poot, G.H.M. Engbers, T. Beugeling, W.G. van Aken, J. Feijen, Improved endothelialization of vascular grafts by local release of growth factor from heparinized collagen matrices. J. Control. Release **64**, 103–114 (2000)

45. G. Bhakta, Z.X.H. Lim, B. Rai, T. Lin, J.H. Hui, G.D. Prestwich, A.J. van Wijnen, V. Nurcombe, S.M. Cool, The influence of collagen and hyaluronan matrices on the delivery and bioactivity of bone morphogenetic protein-2 and ectopic bone formation. Acta Biomater. **9**, 9098–9106 (2013)

46. R. Groning, C. Cloer, R.S. Muller, Development and in vitro evaluation of expandable gastroretentive dosage forms based on compressed collagen sponges. Pharmazie **61**, 608–612 (2006)

47. K.M. Ivester, S.B. Adams, G.E. Moore, D.C. Van Sickle, T.B. Lescun, Gentamicin concentrations in synovial fluid obtained from the tarsocrural joints of horses after implantation of gentamicin-impregnated collagen sponges. Am. J. Vet. Res. **67**, 1519–1526 (2006)

48. G. Ramanathan, S. Thyagarajan, U.T. Sivagnanam, Accelerated wound healing and its promoting effects of biomimetic collagen matrices with siderophore loaded gelatin microspheres in tissue engineering. Mater. Sci. Eng. C **93**, 455–464 (2018)

49. I.V. Yannas, J.F. Burke, C. Huang, P.L. Gordon, Correlation of in vivo collagen degradation rate with in vitro measurements. J. Biomed. Mater. Res. **9**, 623–628 (1975)

50. A. Francesko, D. Soares da Costa, R.L. Reis, I. Pashkuleva, T. Tzanov, Functional biopolymer-based matrices for modulation of chronic wound enzyme activities. Acta Biomater. **9**, 5216–5225 (2013)

51. G.T. Tihan, I. Răuţ, R.G. Zgârian, M.V. Ghica, Collagen-based biomaterials for ibuprofen delivery. Comptes Rendus Chimie **19**, 390–394 (2016)

52. S.H. De Paoli Lacerda, B. Ingber, N. Rosenzweig, Structure–release rate correlation in collagen gels containing fluorescent drug analog. Biomaterials **26**, 7164–7172 (2005)

53. J.L. Holloway, H. Ma, R. Rai, J.A. Burdick, Modulating hydrogel crosslink density and degradation to control bone morphogenetic protein delivery and in vivo bone formation. J. Control. Release **191**, 63–70 (2014)

54. I. Metzmacher, F. Radu, M. Bause, P. Knabner, W. Friess, A model describing the effect of enzymatic degradation on drug release from collagen minirods. Eur. J. Pharm. Biopharm. **67**, 349–360 (2007)

55. M.S. Ågren, C.J. Taplin, J.F. Woessner, W.H. Eagistein, P.M. Mertz, Collagenase in wound healing: Effect of wound age and type. J. Investig. Dermatol. **99**, 709–714 (1992)

56. G. Koopmans, B. Hasse, N. Sinis, Chapter 19 The role of collagen in peripheral nerve repair, in *International Review of Neurobiology*, (Academic Press, 2009), pp. 363–379

57. V. Ferraro, M. Anton, V. Santé-Lhoutellier, The "sisters" α-helices of collagen, elastin and keratin recovered from animal by-products: Functionality, bioactivity and trends of application. Trends Food Sci. Technol. **51**, 65–75 (2016)

58. J. Choi, H. Park, T. Kim, Y. Jeong, M.H. Oh, T. Hyeon, A.A. Gilad, K.H. Lee, Engineered collagen hydrogels for the sustained release of biomolecules and imaging agents: Promoting the growth of human gingival cells. Int. J. Nanomedicine **9**, 5189–5201 (2014)

59. L.M. Mullen, S.M. Best, R.A. Brooks, S. Ghose, J.H. Gwynne, J. Wardale, N. Rushton, R.E. Cameron, Binding and release characteristics of insulin-like growth factor-1 from a collagen–glycosaminoglycan scaffold, tissue engineering. Part C, Methods **16**, 1439–1448 (2010)

60. H.S. Nanda, S. Chen, Q. Zhang, N. Kawazoe, G. Chen, Collagen scaffolds with controlled insulin release and controlled pore structure for cartilage tissue engineering. Biomed. Res. Int. **2014**, 623805 (2014)

61. I. d'Angelo, O. Oliviero, F. Ungaro, F. Quaglia, P.A. Netti, Engineering strategies to control vascular endothelial growth factor stability and levels in a collagen matrix for angiogenesis: The role of heparin sodium salt and the PLGA-based microsphere approach. Acta Biomater. **9**, 7389–7398 (2013)

62. B. Wen, M. Karl, D. Pendrys, D. Shafer, M. Freilich, L. Kuhn, An evaluation of BMP-2 delivery from scaffolds with miniaturized dental implants in a novel rat mandible model. J. Biomed. Mater. Res. B Appl. Biomater. **97B**, 315–326 (2011)

63. T. Kawashima, N. Nagai, H. Kaji, N. Kumasaka, H. Onami, Y. Ishikawa, N. Osumi, M. Nishizawa, T. Abe, A scalable controlled-release device for transscleral drug delivery to the retina. Biomaterials **32**, 1950–1956 (2011)

64. N. Nagai, N. Kumasaka, T. Kawashima, H. Kaji, M. Nishizawa, T. Abe, Preparation and characterization of collagen microspheres for sustained release of VEGF. J. Mater. Sci. Mater. Med. **21**, 1891–1898 (2010)

65. J.M. Wu, Y.Y. Xu, Z.H. Li, X.Y. Yuan, P.F. Wang, X.Z. Zhang, Y.Q. Liu, J. Guan, Y. Guo, R.X. Li, H. Zhang, Heparin-functionalized collagen matrices with controlled release of basic fibroblast growth factor. J. Mater. Sci. Mater. Med. **22**, 107–114 (2011)

66. G. Frenning, Modelling drug release from inert matrix systems: From moving-boundary to continuous-field descriptions. Int. J. Pharm. **418**, 88–99 (2011)

67. N.A. Peppas, B. Narasimhan, Mathematical models in drug delivery: How modeling has shaped the way we design new drug delivery systems. J. Control. Release **190**, 75–81 (2014)

68. R.V. Joshi, C.E. Nelson, K.M. Poole, M.C. Skala, C.L. Duvall, Dual pH- and temperature-responsive microparticles for protein delivery to ischemic tissues. Acta Biomater. **9**, 6526–6534 (2013)

69. M. Ye, S. Kim, K. Park, Issues in long-term protein delivery using biodegradable microparticles. J. Control. Rel. **146**, 241–260 (2010)

70. M.V.S. Varma, A.M. Kaushal, A. Garg, S. Garg, Factors affecting mechanism and kinetics of drug release from matrix-based oral controlled drug delivery systems. Am. J. Drug Deliv. **2**, 43–57 (2004)

71. S. Dash, P.N. Murthy, L. Nath, P. Chowdhury, Kinetic modeling on drug release from controlled drug delivery systems. Acta Pol. Pharm. **67**, 217–223 (2010)

72. M. Otsuka, H. Nakagawa, A. Ito, W.I. Higuchi, Effect of geometrical structure on drug release rate of a three-dimensionally perforated porous apatite/collagen composite cement. J. Pharm. Sci. **99**, 286–292 (2010)

73. R. Gurny, E. Doelker, N.A. Peppas, Modelling of sustained release of water-soluble drugs from porous, hydrophobic polymers. Biomaterials **3**, 27–32 (1982)

74. N.A. Peppas, A model of dissolution-controlled solute release from porous drug delivery polymeric systems. J. Biomed. Mater. Res. **17**, 1079–1087 (1983)

75. R.W. Korsmeyer, R. Gurny, E. Doelker, P. Buri, N.A. Peppas, Mechanisms of potassium chloride release from compressed, hydrophilic, polymeric matrices: Effect of entrapped air. J. Pharm. Sci. **72**, 1189–1191 (1983)

76. N.A. Peppas, R. Gurny, Relation between the structure of polymers and the controlled release of active ingredients. Pharm. Acta Helv. **58**, 2–8 (1983)

77. T. Higuchi, Rate of release of medicaments from ointment bases containing drugs in suspension. J. Pharm. Sci. **50**, 874–875 (1961)

78. P.L. Ritger, N.A. Peppas, A simple equation for description of solute release I. Fickian and non-fickian release from non-swellable devices in the form of slabs, spheres, cylinders or discs. J. Control. Release **5**, 23–36 (1987)

79. R.W. Korsmeyer, R. Gurny, E. Doelker, P. Buri, N.A. Peppas, Mechanisms of solute release from porous hydrophilic polymers. Int. J. Pharm. **15**, 25–35 (1983)

80. N.A. Peppas, J.J. Sahlin, A simple equation for the description of solute release. III. Coupling of diffusion and relaxation. Int. J. Pharm. **57**, 169–172 (1989)

81. C. Mircioiu, V. Voicu, V. Anuta, A. Tudose, C. Celia, D. Paolino, M. Fresta, R. Sandulovici, I. Mircioiu, Mathematical modeling of release kinetics from supramolecular drug delivery systems. Pharmaceutics **11**, 140 (2019)

82. V. Papadopoulou, K. Kosmidis, M. Vlachou, P. Macheras, On the use of the Weibull function for the discernment of drug release mechanisms. Int. J. Pharm. **309**, 44–50 (2006)

83. J. Siepmann, N.A. Peppas, Modeling of drug release from delivery systems based on hydroxypropyl methylcellulose (HPMC). Adv. Drug Deliv. Rev. **48**, 139–157 (2001)

84. K. Kosmidis, E. Rinaki, P. Argyrakis, P. Macheras, Analysis of Case II drug transport with radial and axial release from cylinders. Int. J. Pharm. **254**, 183–188 (2003)

85. A. Dokoumetzidis, P. Macheras, A century of dissolution research: From Noyes and Whitney to the biopharmaceutics classification system. Int. J. Pharm. **321**, 1–11 (2006)

86. S. Giovagnoli, T. Tsai, P.P. DeLuca, Formulation and release behavior of doxycycline–alginate hydrogel microparticles embedded into pluronic F127 thermogels as a potential new vehicle for doxycycline intradermal sustained delivery. AAPS PharmSciTech **11**, 212–220 (2010)

87. A. Dokoumetzidis, P. Macheras, IVIVC of controlled release formulations: Physiological–dynamical reasons for their failure. J. Control. Release **129**, 76–78 (2008)

88. K. Adibkia, M.R.S. Shadbad, A. Nokhodchi, A. Javadzadeh, M. Barzegar-Jalali, J. Barar, G. Mohammadi, Y. Omidi, Piroxicam nanoparticles for ocular delivery: Physicochemical characterization and implementation in endotoxin-induced uveitis. J. Drug Target. **15**, 407–416 (2007)

89. B. Li, K.V. Brown, J.C. Wenke, S.A. Guelcher, Sustained release of vancomycin from polyurethane scaffolds inhibits infection of bone wounds in a rat femoral segmental defect model. J. Control. Release **145**, 221–230 (2010)

90. U. Gbureck, E. Vorndran, J.E. Barralet, Modeling vancomycin release kinetics from microporous calcium phosphate ceramics comparing static and dynamic immersion conditions. Acta Biomater. **4**, 1480–1486 (2008)

91. C.E. Nelson, M.K. Gupta, E.J. Adolph, J.M. Shannon, S.A. Guelcher, C.L. Duvall, Sustained local delivery of siRNA from an injectable scaffold. Biomaterials **33**, 1154–1161 (2012)

92. A. Karewicz, K. Zasada, K. Szczubiałka, S. Zapotoczny, R. Lach, M. Nowakowska, "Smart" alginate–hydroxypropylcellulose microbeads for controlled release of heparin. Int. J. Pharm. **385**, 163–169 (2010)

93. J. Forsgren, E. Jämstorp, S. Bredenberg, H. Engqvist, M. Strømme, A ceramic drug delivery vehicle for oral administration of highly potent opioids. J. Pharm. Sci. **99**, 219–226 (2010)

94. S. Hesaraki, F. Moztarzadeh, R. Nemati, N. Nezafati, Preparation and characterization of calcium sulfate–biomimetic apatite nanocomposites for controlled release of antibiotics. J. Biomed. Mater. Res. B Appl. Biomater. **91B**, 651–661 (2009)

95. H. Liu, C. Wang, Q. Gao, J. Chen, B. Ren, X. Liu, Z. Tong, Facile fabrication of well-defined hydrogel beads with magnetic nanocomposite shells. Int. J. Pharm. **376**, 92–98 (2009)

96. S. Hesaraki, R. Nemati, Cephalexin-loaded injectable macroporous calcium phosphate bone cement. J Biomed Mater Res B Appl Biomater **89B**, 342–352 (2009)

97. A.S. da Silva, C.E. da Rosa Silva, F.R. Paula, F.E.B. da Silva, Discriminative dissolution method for benzoyl metronidazole oral suspension. AAPS PharmSciTech **17**, 778–786 (2016)

98. A. Azadi, M. Hamidi, M.-R. Rouini, Methotrexate-loaded chitosan nanogels as 'Trojan Horses' for drug delivery to brain: Preparation and in vitro/in vivo characterization. Int. J. Biol. Macromol. **62**, 523–530 (2013)

99. A.J. Singer, R.A.F. Clark, Cutaneous wound healing. N. Engl. J. Med. **341**, 738–746 (1999)

100. Y. Tabata et al., Controlled release of vascular endothelial growth factor by use of collagen hydrogels. J. Biomater. Sci., Polymer Edition **11**(9), 915–930 (2000).

101. M. Monaghan et al., A collagen-based scaffold delivering exogenous microrna-29b to modulate extracellular matrix remodeling. Mol. Ther. 22(4), 786–796 (2014).
102. M.J.B. Wissink et al., Immobilization of heparin to EDC/NHS-crosslinked collagen. Characterization and in vitro evaluation. Biomaterials 22(2), 151–163 (2001).
103. Y. Agban et al., Nanoparticle cross-linked collagen shields for sustained delivery of pilocarpine hydrochloride. Int. J. Pharm. 501(1–2), 96–101 (2016).
104. R. Sripriya et al., Collagen bilayer dressing with ciprofloxacin, an effective system for infected wound healing. J. Biomater. Sci. Polym. Ed. 18(3), 335–351 (2007).
105. H. Bentz, J.A. Schroeder, T.D. Estridge, Improved local delivery of TGF-beta2 by binding to injectable fibrillar collagen via difunctional polyethylene glycol. J. Biomed. Mater. Res. 39(4), 539–548 (1998).
106. S. Koch et al., Enhancing angiogenesis in collagen matrices by covalent incorporation of VEGF. J. Mater. Sci. Mater. Med. 17(8), 735–741 (2006).
107. M. Ragothaman, T. Palanisamy, C. Kalirajan, Collagen–poly(dialdehyde) guar gum based porous 3D scaffolds immobilized with growth factor for tissue engineering applications. Carbohydr. Polym. **114**, 399–406 (2014).
108. M. Markowicz et al., Effects of modified collagen matrices on human umbilical vein endothelial cells. Int J. Artif. Organs 28(12), 1251–1258 (2005).
109. E.O. Osidak et al., Regulation of the binding of the BMP-2 growth factor with collagen by blood plasma fibronectin. Appl. Biochem. Microbiol. 50(2), 200–205 (2014).
110. H.S. Nanda et al., Preparation of collagen porous scaffolds with controlled and sustained release of bioactive insulin. J. Bioactive Compatib. Polymers Biomed. Appl. 29(2), 95–109 (2014).
111. E. Quinlan et al., Development of collagen–hydroxyapatite scaffolds incorporating PLGA and alginate microparticles for the controlled delivery of rhBMP-2 for bone tissue engineering. J. Control. Release **198**, 71–79 (2015).
112. B. Wen et al., An evaluation of BMP-2 delivery from scaffolds with miniaturized dental implants in a novel rat mandible model. J. Biomed. Mater. Res. B Appl. Biomater. 97(2), 315–326 (2011).
113. P. Prabu et al., Preparation and drug release activity of scaffolds containing collagen and poly(caprolactone). J. Biomed. Mater. Res. A 79A(1), 153–158 (2006).
114. N. Shanmugasundaram et al., Design and delivery of silver sulfadiazine from alginate microspheres-impregnated collagen scaffold. J. Biomed. Mater. Res. B Appl. Biomater. 77B(2), 378–388 (2006).
115. M. Schlapp, W. Friess, Collagen/PLGA microparticle composites for local controlled delivery of gentamicin. J. Pharm Sci. 92(11), 2145–2151 (2003).
116. A.L. Weiner et al., Liposome–collagen gel matrix: A novel sustained drug delivery system. J. Pharma. Sci. 74(9), 922–925 (1985).
117. T. Kitajima, H. Terai, Y. Ito, A fusion protein of hepatocyte growth factor for immobilization to collagen. Biomaterials 28(11), 1989–1997 (2007).
118. E. Jeon et al., Engineering and application of collagen-binding fibroblast growth factor 2 for sustained release. J. Biomed. Mater. Res. A 102(1), 1–7 (2014).
119. W. Sun et al., The effect of collagen-binding NGF-β on the promotion of sciatic nerve regeneration in a rat sciatic nerve crush injury model. Biomaterials 30(27), 4649–4656 (2009).
120. Y. Yang et al., Collagen-binding human epidermal growth factor promotes cellularization of collagen scaffolds. Tissue Eng Part A 15(11), 3589–3596 (2009).
121. J. Zhang et al., Collagen-targeting vascular endothelial growth factor improves cardiac performance after myocardial infarction. Circulation 119(13), 1776–1784 (2009).
122. N. Ohkawara et al., Hepatocyte growth factor fusion protein having Collagen-Binding Activity (CBD-HGF) accelerates re-endothelialization and intimal hyperplasia in balloon-injured rat carotid artery. J. Atheroscler. Thromb. 14(4), 185–191 (2007).
123. W. Sun et al., Collagen scaffolds loaded with collagen-binding NGF-beta accelerate ulcer healing. J. Biomed. Mater. Res. A 92(3), 887–895 (2010).

124. C. Shi et al., Regeneration of full-thickness abdominal wall defects in rats using collagen scaffolds loaded with collagen-binding basic fibroblast growth factor. Biomaterials 32(3), 753–759 (2011).
125. M.A. Urello, K.L. Kiick, M.O. Sullivan, A CMP-based method for tunable, cell-mediated gene delivery from collagen scaffolds. J. Mater. Chem. B 2(46), 8174–8185 (2014).
126. T. Takezawa et al., Collagen vitrigel membrane useful for paracrine assays in vitro and drug delivery systems in vivo. J. Biotechnol. 131(1), 76–83 (2007).
127. J. Zhao et al., Bone regeneration using collagen type I vitrigel with bone morphogenetic protein-2. J. Biosci. Bioeng. 107(3), 318–323 (2009).
128. D. Gopinath et al., Pexiganan-incorporated collagen matrices for infected wound-healing processes in rat. J. Biomed. Mater. Res. Part A 73A(3), 320–331 (2005).
129. T. Kawashima et al., A scalable controlled-release device for transscleral drug delivery to the retina. Biomaterials 32(7), 1950–1956 (2011).
130. M.-A. Lauzon, B. Marcos, N. Faucheux, Effect of initial pBMP-9 loading and collagen concentration on the kinetics of peptide release and a mathematical model of the delivery system. J. Control. Release **182**, 73–82 (2014).

Chapter 4
Collagen Biografts for Chronic Wound Healing

4.1 Introduction

A drug delivery system needs to be "tunable" to achieve a broad range of release profiles, meaning it should be able to modulate the timing, duration, dosage, and location of drug release. A tunable drug delivery system can provide several advantages over conventional routes of administration, enabling release over days, weeks, or months after implantation/injection. This requires the drug delivery system to be capable of providing a broad range of release profiles, based on the application demand. For example, in an application requiring healing of acute wounds, a rapid initial release followed by a more sustained release may be needed for the drug to quickly reach an effective therapeutic concentration and extend release for required healing in about 8–12 weeks [1]. On the other hand, in an application requiring treatment of chronic wounds, more sustained release profiles might be required to prolong drug release beyond 12 weeks [2].

From the wound healing perspective, an ideal drug delivery system should allow noninvasive, repeatable, and reliable switching of therapeutic agent flux [3]. It should provide the structural and mechanical support for the cellular infiltration and growth while promoting the neovascularization at a wound site. Cell adhesion and biological signaling response must be induced, and the physical, chemical, and degradation characteristics of the material must be tunable. A greater challenge is to deliver more than one active reagent in a controlled manner using a single platform. For instance, in tissue engineering applications such as vasculogenesis, there is often a need to provide spatiotemporal delivery of two growth factors, requiring a complex engineering design for a drug delivery vehicle [4]. Although designing such controlled release systems from collagen may appear challenging and costly, careful engineering of material and choice of an appropriate method for immobilizing active agents in collagen can offer a solution.

© The Author(s), under exclusive license to Springer Nature Switzerland AG 2021   53
R. Joshi, *Collagen Biografts for Tunable Drug Delivery*, SpringerBriefs in
Applied Sciences and Technology, https://doi.org/10.1007/978-3-030-63817-7_4

In the following sections, we first look at the pathophysiology of chronic wounds compared to normal wounds and then review some of the commercially available collagen-based wound dressings, their advantages, and their limitations.

4.2   Application of Collagen for Wound Healing

Chronic wounds are defined by the presence of a skin defect or lesion that persists longer than 6 weeks or has a frequent recurrence [5]. Chronic wounds affect around 6.5 million patients in the United States alone [6], and as many as 37 million globally [7]. The fact that an excess of US$25 billion is spent annually on the treatment of chronic wounds indicates how chronic wounds are posing a tremendous burden to the patients' health as well as the economic system. The burden is escalating more and more due to increasing health care costs, an aging population, and a higher incidence of diabetes and obesity [8].

A common yet seriously challenging example of chronic wounds is a diabetic foot ulcer (DFU) which due to its suboptimal healing properties increases the risk of infection, and if not cured promptly leads to leg amputation [9]. In 2010, about 73,000 non-traumatic lower-limb amputations were performed due to DFUs [10]. Costing $38,077 per amputation procedure, approximately 3 billion dollars are spent per year on diabetes-related amputations [11]. An estimated 12% of individuals with a foot ulcer require foot amputation, which is a serious concern because the 5-year survival rate after one major lower extremity amputation is about 50% [8].

Chronic wounds fail to heal because of the disruption of the orderly sequence of events during the wound healing process. To understand the pathophysiology of chronic wounds, it is necessary to know the physiology of the normal wound healing process first. Wound healing normally involves a complex interaction between epidermal and dermal cells, the extracellular matrix (ECM), angiogenesis, and plasma-derived proteins, all coordinated through an array of cytokines and growth factors. This dynamic process can be classified into four overlapping phases, including hemostasis, inflammation, proliferation, and remodeling [12, 13], as depicted in Fig. 4.1 and described briefly below.

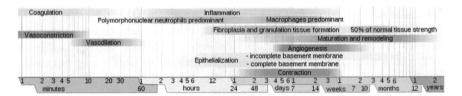

Fig. 4.1 Phases and timeline of wound healing. (The figure is adapted from Häggström, M (2014). "Medical gallery of Mikael Häggström 2014". *WikiJournal of Medicine* 1 (2). https://doi. org/10.15347/wjm/2014.008. ISSN 2002-4436. Creative Commons CC0 1.0 Universal Public Domain Dedication. Used with permission)

(i) *Hemostasis:* After tissue injury, thrombus formation requires interaction between endothelial cells, platelets, and coagulation factors to achieve hemostasis. Trapped platelets within the clot trigger an inflammatory response through the release of vasodilators, chemoattractants, and activation of the complement cascade.

(ii) *Inflammation:* In the early phase of inflammation, neutrophils predominate and remove bacteria and other foreign material from the wound by phagocytosis and release of enzymes. Later in the inflammatory phase neutrophils reduce in number and are replaced by macrophages. This stage lasts until about 48 h after injury.

(iii) *Proliferation:* In this phase, fibroblasts play an important role in the synthesis of new type I collagen and ECM. Additionally, tenascin, fibronectin, and proteoglycans are also produced. Production of ECM is clinically seen as the formation of granulation tissue. The formation of new tissue combined with the contraction of surrounding tissues is essential for the healing of wounds. While a new matrix is synthesized, the existing matrix in and around the wound margin is degraded by several enzyme systems such as matrix metalloproteinases (MMPs) and plasminogen activators. This stage occurs about 2–10 days after injury.

(iv) *Remodeling:* In this phase, type I collagen replaces fibronectin becoming the predominant ECM constituent and resulting in a more mature ECM. Once closure of the wound has been achieved, remodeling of the resulting scar occurs over months or years, with a reduction of cell content and blood flow in the scar tissue.

Scarring and fibrosis in a wound are typically affected by Granzyme B activity (Fig. 4.2). Decorin, for example, is thought to be predominately an antifibrotic proteoglycan and may inhibit excessive collagen production by fibroblasts. As decorin is critical for proper collagen organization and tensile strength, Granzyme B-mediated decorin degradation would result in collagen spacing and organizational defects that lead to reduced tensile strength [14]. Numerous studies have shown that decorin is reduced in fibrotic tissues in the skin and other organs, leading to the disorganized collagen that characterizes these lesions.

An important feature of the proliferation phase in normal wound healing is neovascularization. The dynamic interactions between endothelial cells, various soluble angiogenic cytokines, and the ECM environment promote neovascularization in the wound are depicted in reviews such as [13, 14]. Angiogenic capillaries sprout and invade the fibrin/fibronectin-rich wound clot and organize into a microvascular network throughout the granulation tissue within a few days.

However, in chronic wounds, this dynamic spatiotemporal interaction between endothelial cells, angiogenesis factors, and surrounding ECM proteins is impaired. The chronic skin defect is usually in a permanent inflammatory state due to a hyperstimulated neutrophil response [5]. Along with an elevated level of pro-inflammatory cytokines, permanent increased proteolytic activity is typical for chronic wounds, contributed by excessive production of matrix metalloproteinases (MMPs) in the

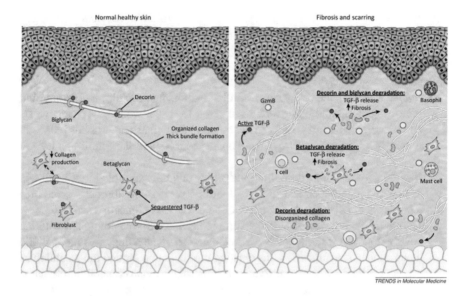

Fig. 4.2 Granzyme B activity in fibrosis and scarring. Degradation of decorin by granzyme B (GzmB) can be profibrotic by promoting disorganized collagen. Fibroblasts may also be encouraged to produce more collagen in the absence of inhibitory signals from the decorin proteoglycan. GzmB-mediated degradation of decorin, biglycan, and beta glycan can also result in the release of active transforming growth factor-β (TGF-β), potentially increasing the fibrotic response. (Figure adapted from Hiebert et al. [14] with permission)

wound [15, 16]. MMPs are said to be responsible for poor healing by breaking down many components of the ECM and by inhibiting growth factors that are essential for tissue synthesis [17]. This imbalance between ECM deposition and degradation, and deficiencies in growth factor and cytokine receptors, lead to impaired progenitor cell recruitment and angiogenesis and delay wound epithelialization [18, 19], as depicted in Fig. 4.3.

Typically, wound debridement followed by its compression with sterile gauze is the classic treatment for treating acute wounds [20, 21]. However, when this method is not effective enough, chronic wounds have to be dressed with adequate biomaterials to protect the long-term healing from infection and aiding in tissue regeneration [21, 22].

An intervention from an alternative multifunctional tissue engineering strategy based on collagen, a molecule native to our body, can offer a great solution for chronic wound healing. Collagen can provide a strong structural template for cell infiltration and growth of new tissue, and at the same time, provide local exposure of growth factors that can coordinate angiogenic response for full functional tissue recovery. The ultimate goal for treating these wounds is scar-free healing and timely restoration of tissue function [23].

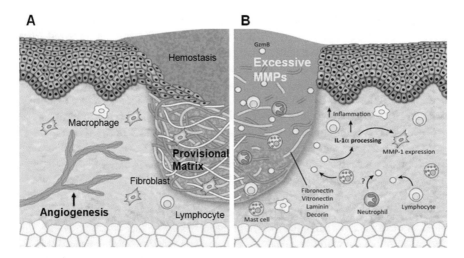

Fig. 4.3 Normal versus impaired wound healing. Normal wound healing (**a**) versus impaired wound healing (**b**). In a normal wound healing, fibroblasts construct new ECM necessary to support cell ingrowth, and blood vessels that carry oxygen and nutrients necessary for cell survival. The provisional ECM promotes granulation tissue formation. Macrophages, fibroblasts, and blood vessels move into the wound space as a unit, through dynamic biologic interactions contributing to tissue repair. Fibroblasts contribute to new type I collagen synthesis. While MMP levels decrease through the normal wound-healing process, chronic wounds continue to show a significantly higher level of proteases and pro-inflammatory cytokines. As a result, inflammation persists longer and higher levels of MMPs cause excessive breakdown of type I collagen and ECM. A chronic wound healing is then further impaired by a lack of neovascularization and an impaired reepithelialization. (The figure is adapted with permission from Hiebert et al. [14])

4.3   Collagen-Based Materials as Wound Dressings Available Commercially

As a major natural constituent of our body, collagen is seen to play an integral role in the repair and replacement of soft tissue by providing an extracellular scaffold, stimulating certain growth factors, and propagating tissue granulation [24]. As a result, numerous efforts have been put into developing collagen implants and wound dressings to specifically accelerate the natural process of wound healing and promote tissue regeneration.

Dhivya et al. [25] point out characteristics of an ideal wound dressing that should be selected based on its ability to (a) provide or maintain a moist environment, (b) enhance epidermal migration, (c) promote angiogenesis and connective tissue synthesis, (d) allow gas exchange between wounded tissue and environment, (e) maintain appropriate tissue temperature to improve the blood flow to the wound bed and enhance epidermal migration, (f) protect against bacterial infection, (g) be nonadherent to the wound and easy to remove after healing, (h) provide debridement action to enhance leucocytes migration and support the accumulation of enzyme, and (i) be sterile, nontoxic, and nonallergic. While various types of scaffolds prepared from synthetic or natural materials have been engineered to meet many of these criteria, their effectiveness remains limited by slow neovascularization ultimately contributing to poor functional integration, pain, and/or scarring. As such, many efforts have been lately focused on the design and development of collagen-based biomaterials that can provide the structural and mechanical support for the cellular infiltration and growth, while promoting the neovascularization at the wound site.

Many advanced wound dressings and skin substitutes have been introduced in the wound care market during the last decades [22, 24, 26–31]. Literature shows that available skin-grafts for wound healing can be classified in different ways. Human skin or dermal equivalent (HSE) has two types of tissue-engineered substitutes available, one mimics the layer of skin composed of Keratinocytes and fibroblast on the collagen matrix (Cell containing matrix). The second contains only the dermal components with fibroblasts on the collagen matrix (Acellular matrix) [25]. The major mechanism of HSE is to secrete and stimulate the wound growth factor by which epithelialization is achieved. Bioengineered substitutes are capable of adapting to their environment so that they can release growth factors and cytokines incorporated in dressings. Bioengineered dressings are suitable for diabetic foot ulcers and venous leg ulcers. Apligraf, the FDA approved skin equivalent substitute consists of keratinocytes and fibroblast-seeded collagen for venous ulcers. Some skin substitutes commercially available include, Alloderm™ composed of normal human fibroblasts with all cellular materials removed and Integra™ artificial skin consists of collagen/chondroitin 6 sulfate matrix overlaid with a thin silicone sheet. Other few substitutes are Laserskin™, Biobrane ™, Bioseed™, and Hyalograft3-DTM.

The constituent scheme of the various skin biografts available commercially is given in Fig. 4.4. Biografts can be classified as permanent, semi-permanent, or temporary based on the duration of cover; or as epidermal, dermal, or dermo-epidermal (composite) based on the anatomical structure [32]. They can also be classified based on skin substitute composition as cellular or acellular. Similarly, they can be biological (autologous, allogeneic, xenogeneic) or synthetic (biodegradable, non-biodegradable) based on the type of formulating biomaterial.

A variety of products have been commercially developed and reviewed in detail by various researchers [24, 26–30, 33]. From a collagen-based biograft perspective selective examples of acellular products, including Oasis, Alloderm, and Integra dermal regeneration template products are described below. It is claimed that the collagen in these products promotes the deposit of newly formed collagen in the wound bed. These dressings come in a variety of formats, including pads, gels, and particle forms. They can be used on surgical wounds, in deep wounds to fill dead space, to absorb, exudate, and provide a moist environment. Below, some advantages and disadvantages of the commercially available biografts are discussed.

Alloderm (TM), distributed by the LifeCell Corporation, is a processed acellular dermal matrix derived from human cadavers [34]. Cadaveric tissue samples are first screened for a host of transmittable pathogens. The decellularization is achieved through the use of a detergent solution that leaves only the dermal matrix and

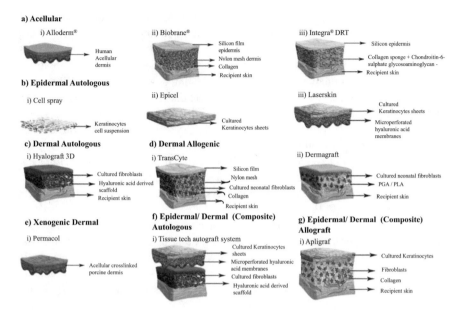

Fig. 4.4   Commercially available tissue-engineered skin substitutes. (a) Acellular: (i) Karoderm (ii) Biobrane (iii) Integra **(b)** Epidermal Autologous: (i) Cell spray (ii) Epicel (iii) Laser skin **(c)** Dermal autologous: (i) Hyalograft 3D **(d)** Dermal allogenic: (i) TransCyte (ii) Dermagraft **(e)** Dermal xenogenic: (i) Permacol **(f)** Epidermal/Dermal (composite) autologous i. Tissue tech autograft system **(g)** Epidermal/Dermal (composite) allograft (i) Apligraf. (Figure adapted from KomalVig et al. [32], doi:https://doi.org/10.3390/ijms18040789,covered under the creative commons license https://creativecommons.org/licenses/by/4.0/)

associated basal lamina intact, removing all other cellular components [35]. Following decellularization, samples are lyophilized for storage and must be rehydrated before use [36]. Upon grafting, host fibroblasts and associated vasculature infiltrate the Alloderm matrix. However, the full extent of vascularization is said to be uncertain [37, 38]. The clinical use of Alloderm also requires subsequent application of an ultrathin split-thickness autograft immediately following implantation, since Alloderm lacks an epidermal component and has limited barrier properties [37]. Other disadvantages of Alloderm are said to be the requirement of multiple applications and a theoretical risk of transmission of human pathogens [39].

Oasis (TM), developed by Cook Biotech, is an acellular dermal scaffold derived from porcine small intestinal submucosa (SIS). It contains numerous dermal components including collagen, glycosaminoglycans (hyaluronic acid), proteoglycans, fibronectin, and bioactive growth factors such as fibroblast growth factor-2, transforming growth factor $\beta1$, and vascular endothelial growth factor (VEGF) present naturally in the SIS [40]. Following application to the wound bed, this acellular matrix is infiltrated by fibroblasts and associated vasculature, which gradually replace the material with new ECM components over time [41]. It should be noted that while the material has a limited porosity, it does not provide a moisture barrier and must be protected by an appropriate secondary dressing [42]. Recently, the bolster technique, a novel method designed to enable the small intestinal submucosa wound matrix (SISWM) (Oasis Wound Matrix, Smith & Nephew, Fort Worth, Texas) was successfully reported to impart its wound healing properties to complex traumatic wounds [43]. The method by which SISWM is applied to a wound is important because it needs to impart the critical wound healing components that will support tissue reconstruction and angiogenesis, attack and disrupt biofilms, and restore normal anatomic integrity to the patient's disrupted dermis and epidermis.

Some of the Oasis limitations may include a possible higher infection rate and the need for multiple applications [39]. However, a clinical trial comparing the application of Oasis in 73 patients with diabetic foot ulcers showed only slight statistical superiority (p = 0.055) when compared with Regranex – a carboxymethyl-cellulose-based topical gel containing recombinant human platelet-derived growth factor (rhPDGF-BB) [44].

Integra (TM) is a composite acellular collagen product developed by the Integra Life Science Corporation. It is composed of an outer layer of silicone and a cross-linked bovine type I collagen glycosaminoglycan dermal matrix and was originally described by Yannas and Burke [45, 46]. The collagen-GAG matrix is gradually invaded by host fibroblasts upon implantation in an excised wound bed [37]. Tissue integration is said to take place in approximately 3–6 weeks, resulting in the production of a "neodermis" with associated vasculature [37]. During this time, the silicone layer acts as a protective barrier, limiting moisture loss through the membrane [35]. Once the dermal layer regenerates, the silicone layer has to be removed and the wound is permanently closed with a split-thickness skin graft (STSG) [24]. However, Singer et al. showed that a successful treatment STSG with a novel protocol modification utilizing Integra Bilayer Matrix Wound Dressing (Integra LifeSciences Corporation, Plainsboro, New Jersey) alone negated the necessity for STSG [47]. This case introduced a novel method of treating avulsions with a lower cost of care, less patient morbidity, and more aesthetically pleasing results.

While Integra has shown promise in the treatment of chronic wounds and burns, it has some limitations that hinder its clinical use. When compared with AlloDerm in a mouse wound model, the Integra matrix-induced more foreign body response and giant cells, because it is a chemically cross-linked material [48]. Integra scaffold needs to be first cleared by macrophages to allow the deposition of collagen fibers. Since Integra has no intrinsic immunological defenses, it can be easily infected by bacteria and requires daily monitoring for signs of bacterial growth until the bio integration process is complete [49]. In the incidence of infection, wound debridement and reapplication are typically necessary, which further lengthens the time required for healing [49]. Another concern is a two-step process required for Integra based therapy. Since the silicone layer of Integra functions only as a temporary covering, it must be replaced by an ultrathin autograft following neodermal formation [27]. Given that the average acceptance rate of Integra is at least 10% lower than for a standard split-thickness graft, patients might prefer to undergo the latter procedure directly instead of opting for a riskier two-step process, if they have sufficient donor skin [27]. Furthermore, technical difficulty in the Integra application necessitates physician training, and as a result, it may only be used by practitioners that have undergone a company-sponsored training program [50]. In an early trial, incidences of hematoma and seroma formation occurred due to improper application of Integra, highlighting the level of skill required for proper use of the material [46].

There are very few commercially available collagen dressings that use a drug delivery mechanism to heal chronic wounds such as diabetic, venous, and pressure ulcers, and burns. These dressings include ColActive® Plus Ag, Promogran Prisma™ Ag, and Biostep Ag, summarized in Table 4.1. It is observed that the majority of these products are restricted to the delivery of an antimicrobial agent, mainly – silver (to prevent infection in chronic wounds), and ethylenediaminetetracetic acid (EDTA) (to form a chelating complex with MMPs and to prevent them from the excessively degrading matrix) [51].

Literature regarding collagen-based wound dressings capable of drug delivery should be cautiously analyzed, because many of these studies may have different surgical protocols, different types and severity of wounds, and different ages of the patients [52]. Fitzgerald et al. have summarized various collagen-based wound dressings, informing readers about the advanced collagen dressing constituents, appropriate indications for each product's usage, mechanism of action of the dressing, and its cost analysis [53].

Table 4.1   Collagen-based wound dressings with drug delivery capability

Drug delivery system	Company	Drug incorporated	Product formulation	Drug incorporation method	References
ColActive® Plus Ag	Covalon	EDTA and silver ions	Lyophilized collagen sponge made with collagen, carboxyl methylcellulose (CMC), and sodium alginate	Immersing collagen sponge in drug solution and drying it	[54, 55]
Promogran Prisma™ Ag	Acelity	Silver-ORC containing 25% w/w ionically bound silver (Ag)	Lyophilized sponge consisting of 44% oxidized regenerated cellulose (ORC), 55% collagen, and 1% silver-ORC	Soaking collagen sponge in a drug solution	[56, 57]
Biostep Ag	Smith and Nephew	EDTA and silver (Ag)	Lyophilized sponge made from porcine type I collagen and gelatin	–	[54, 55]
Collatamp G	Innnocol	Gentamicin	Lyophilized sponge created by freeze-drying dispersion of insoluble fibrillar collagen or isolated collagen mixed with a drug solution, and dehydrated	Admixing followed by lyophilization	[58–60]

Overall, looking at the various collagen-based advanced wound dressings, it is seen that despite some advantages, undesirable outcomes may still limit the use of these products in the treatment of chronic wound ulcers. In general, peripheral ischemia, which is a pathological characteristic of chronic ulcers, critically affects collagen-based biomaterial transplantations [61, 62]. Many diabetic patients need surgical revascularization to achieve timely and durable healing. However, with collagen-based wound therapies, it currently takes 3–4 weeks for engineered dermal substitutes to be sufficiently vascularized before a split-thickness skin graft can be placed on the neodermis [63]. Thus, slow vascularization along with the inability of collagen-based dressings to serve as stand-alone therapy adds to the current limitations of collagen-based wound healing products, including frequent surgical interventions, high costs of treatment, and inflammation-mediated response that leads to scar formation rather than tissue regeneration [24, 26–30]. A scope still exists for developing a commercially viable collagen-based drug delivery device that provides an improved vascularization in a shorter period, while promoting tissue regeneration and integration at the site of injection/implantation.

References

1. N.J. Percival, Classification of wounds and their management. Surgery (Oxford) **20**, 114–117 (2002)
2. K.G. Harding, H.L. Morris, G.K. Patel, Healing chronic wounds. BMJ **324** (2002)
3. B.P. Timko, D.S. Kohane, Drug-delivery systems for tunable and localized drug release, Israel J. Chem., (2013) n/a-n/a.
4. A.S. Fu, T.R. Thatiparti, G.M. Saidel, H.A. von Recum, Experimental studies and modeling of drug release from a tunable affinity-based drug delivery platform. Ann. Biomed. Eng. **39**, 2466–2475 (2011)
5. T. Wild, A. Rahbarnia, M. Kellner, L. Sobotka, T. Eberlein, Basics in nutrition and wound healing. Nutrition **26**, 862–866 (2010)
6. A.J. Singer, R.A.F. Clark, Cutaneous wound healing. N. Engl. J. Med. **341**, 738–746 (1999)
7. S. Tejiram, S.L. Kavalukas, J.W. Shupp, A. Barbul, 1 – Wound healing A2 – Ågren, in *Wound Healing Biomaterials*, ed. by S. Magnus, (Woodhead Publishing, 2016), pp. 3–39
8. C.K. Sen, G.M. Gordillo, S. Roy, R. Kirsner, L. Lambert, T.K. Hunt, F. Gottrup, G.C. Gurtner, M.T. Longaker, Human skin wounds: A major and snowballing threat to public health and the economy. Wound Repair Regen. **17**, 763–771 (2009)
9. A. Alavi, R.G. Sibbald, D. Mayer, L. Goodman, M. Botros, D.G. Armstrong, K. Woo, T. Boeni, E.A. Ayello, R.S. Kirsner, Diabetic foot ulcers: Part I. Pathophysiology and prevention. J. Am. Acad. Dermatol. **70**, 1.e1–1.e18 (2014)
10. C.f.D.C.a. Prevention, Centers for Disease Control and Prevention. National Diabetes Statistics Report: Estimates of Diabetes and Its Burden in the United States, 2014, in Atlanta, GA (2014)
11. A. Shearer, P. Scuffham, A. Gordois, A. Oglesby, Predicted costs and outcomes from reduced vibration detection in people with diabetes in the U.S. Diabetes Care **26**, 2305 (2003)
12. K.G. Harding, H.L. Morris, G.K. Patel, Science, medicine and the future: Healing chronic wounds. BMJ **324**, 160–163 (2002)
13. G.C. Gurtner, S. Werner, Y. Barrandon, M.T. Longaker, Wound repair and regeneration. Nature **453**, 314–321 (2008)

14. P.R. Hiebert, D.J. Granville, Granzyme B in injury, inflammation, and repair. Trends Mol. Med. **18**, 732–741 (2012)

15. E.A. Rayment, Z. Upton, G.K. Shooter, Increased matrix metalloproteinase-9 (MMP-9) activity observed in chronic wound fluid is related to the clinical severity of the ulcer. Br. J. Dermatol. **158**, 951–961 (2008)

16. D.R. Yager, B.C. Nwomeh, The proteolytic environment of chronic wounds. Wound Repair Regen. **7**, 433–441 (1999)

17. M. Muller, C. Trocme, B. Lardy, F. Morel, S. Halimi, P.Y. Benhamou, Matrix metalloproteinases and diabetic foot ulcers: The ratio of MMP-1 to TIMP-1 is a predictor of wound healing. Diabet. Med. **25**, 419–426 (2008)

18. T.N. Demidova-Rice, J.T. Durham, I.M. Herman, Wound healing angiogenesis: Innovations and challenges in acute and chronic wound healing. Adv. Wound Care **1**, 17–22 (2012)

19. R. Nunan, K.G. Harding, P. Martin, Clinical challenges of chronic wounds: Searching for an optimal animal model to recapitulate their complexity. Dis. Model. Mech. **7**, 1205–1213 (2014)

20. J.G. Powers, L.M. Morton, T.J. Phillips, Dressings for chronic wounds. Dermatol. Ther. **26**, 197–206 (2013)

21. J.S. Boateng, K.H. Matthews, H.N.E. Stevens, G.M. Eccleston, Wound healing dressings and drug delivery systems: A review. J. Pharm. Sci. **97**, 2892–2923 (2008)

22. L.I.F. Moura, A.M.A. Dias, E. Carvalho, H.C. de Sousa, Recent advances on the development of wound dressings for diabetic foot ulcer treatment—A review. Acta Biomater. **9**, 7093–7114 (2013)

23. A.S. Halim, T.L. Khoo, S.J.M. Yussof, Biologic and synthetic skin substitutes: An overview. Indian J. Plast. Surg. **43**, S23–S28 (2010)

24. K.S. Vyas, H.C. Vasconez, Wound healing: Biologics, skin substitutes, biomembranes and scaffolds. Healthcare **2**, 356–400 (2014)

25. S. Dhivya, V.V. Padma, E. Santhini, Wound dressings – a review. Biomedicine (Taipei) **5**, 22–22 (2015)

26. L. Yazdanpanah, M. Nasiri, S. Adarvishi, Literature review on the management of diabetic foot ulcer. World J. Diabetes **6**, 37–53 (2015)

27. C. Hrabchak, L. Flynn, K.A. Woodhouse, Biological skin substitutes for wound cover and closure. Expert Rev Med Devices **3**, 373–385 (2006)

28. Y.M. Bello, A.F. Falabella, W.H. Eaglstein, Tissue-engineered skin. Current status in wound healing. Am. J. Clin. Dermatol. **2**, 305–313 (2001)

29. R.A. Kamel, J.F. Ong, E. Eriksson, J.P.E. Junker, E.J. Caterson, Tissue engineering of skin. J. Am. Coll. Surg. **217**, 533–555 (2013)

30. T. Biedermann, S. Boettcher-Haberzeth, E. Reichmann, Tissue engineering of skin for wound coverage. Eur. J. Pediatr. Surg. **23**, 375–382 (2013)

31. R.G. Frykberg, T. Zgonis, D.G. Armstrong, V.R. Driver, J.M. Giurini, S.R. Kravitz, A.S. Landsman, L.A. Lavery, J.C. Moore, J.M. Schuberth, D.K. Wukich, C. Andersen, J.V. Vanore, Diabetic foot disorders: A clinical practice guideline (2006 Revision). J. Foot Ankle Surg. **45**, S1–S66 (2006)

32. K. Vig, A. Chaudhari, S. Tripathi, S. Dixit, R. Sahu, S. Pillai, V.A. Dennis, S.R. Singh, Advances in skin regeneration using tissue engineering. Int. J. Mol. Sci. **18**, 789 (2017)

33. J.N. Kearney, Clinical evaluation of skin substitutes. Burns **27**, 545–551 (2001)

34. F.A. Auger, F. Berthod, V. Moulin, R. Pouliot, L. Germain, Tissue-engineered skin substitutes: From in vitro constructs to in vivo applications. Biotechnol. Appl. Biochem. **39**, 263–275 (2004)

35. M. Balasubramani, T.R. Kumar, M. Babu, Skin substitutes: A review. Burns **27**, 534–544 (2001)

36. R.J. Snyder, Treatment of nonhealing ulcers with allografts. Clin. Dermatol. **23**, 388–395 (2005)

37. I. Jones, L. Currie, R. Martin, A guide to biological skin substitutes. Br. J. Plast. Surg. **55**, 185–193 (2002)
38. P.G. Shakespeare, The role of skin substitutes in the treatment of burn injuries. Clin. Dermatol. **23**, 413–418 (2005)
39. H. Cronin, G. Goldstein, Biologic skin substitutes and their applications in dermatology. Dermatol. Surg. **39**, 30–34 (2013)
40. E.N. Mostow, G.D. Haraway, M. Dalsing, J.P. Hodde, D. King, Effectiveness of an extracellular matrix graft (OASIS Wound Matrix) in the treatment of chronic leg ulcers: A randomized clinical trial. J. Vasc. Surg. **41**, 837–843 (2005)
41. M. Brown-Etris, W.D. Cutshall, M.C. Hiles, A new biomaterial derived from small intestine submucosa and developed into a wound matrix device. Wounds Compendium Clin. Res. Pract. **14**, 150–166 (2002)
42. R.A. Santucci, T.D. Barber, Resorbable extracellular matrix grafts in urologic reconstruction. Int. Braz. J. Urol. **31**, 192–203 (2005)
43. F.J. Collini, S.C. Stevenson, J.P. Hodde, The bolster technique utilising small intestinal submucosa wound matrix: A novel approach to wound treatment. Int. Wound J. **16**, 1222–1229 (2019)
44. J.A. Niezgoda, C.C. Van Gils, R.G. Frykberg, J.P. Hodde, Randomized clinical trial comparing OASIS wound matrix to Regranex Gel for diabetic ulcers. Adv. Skin Wound Care **18**, 258–266 (2005)
45. I. Yannas, J. Burke, D. Orgill, E. Skrabut, Wound tissue can utilize a polymeric template to synthesize a functional extension of skin. Science **215**, 174–176 (1982)
46. J.F. Burke, I.V. Yannas, W.C. Quinby, C.C. Bondoc, W.K. Jung, Successful use of a physiologically acceptable artificial skin in the treatment of extensive burn injury. Ann. Surg. **194**, 413–428 (1981)
47. M. Singer, J. Korsh, W. Predun, D. Warfield Jr., R. Huynh, T. Davenport, L. Riina, A novel use of integra™ bilayer matrix wound dressing on a pediatric scalp avulsion: A case report. Eplasty **15**, e8–e8 (2015)
48. H. Debels, M. Hamdi, K. Abberton, W. Morrison, Dermal matrices and bioengineered skin substitutes: A critical review of current options. Plast. Reconstr. Surg. Glob. Open **3**, e284 (2015)
49. J.T. Schulz 3rd, R.G. Tompkins, J.F. Burke, Artificial skin. Annu. Rev. Med. **51**, 231–244 (2000)
50. K.H. Lee, Tissue-engineered human living skin substitutes: Development and clinical application. Yonsei Med. J. **41**, 774–779 (2000)
51. D. Silcock, Collagen-based dressings as therapeutic agents for wound healing, in *Drug-device combination products – Delivery technologies and applications*, ed. by L. Andrew, (Woodhead Publishing, 2010), pp. 280–310
52. C. Holmes, J.S. Wrobel, M.P. Maceachern, B.R. Boles, Collagen-based wound dressings for the treatment of diabetes-related foot ulcers: A systematic review. Diabetes Metab. Syndr. Obes. **6**, 17–29 (2013)
53. R.H. Fitzgerald, J.S. Steinberg, Collagen in wound healing: Are we onto something new or just repeating the past? Foot Ankle Online J. **2** (2009)
54. V. DiTizio, F. DiCosmo, Y. Xiao, Non-adhesive elastic gelatin matrices, in Google Patents (2013)
55. http://www.covalon.com/advanced-wound-care-products, in Covalon Website
56. B.M. Cullen, D.W. Silcock, Wound dressing compositions comprising chitosan and an oxidised cellulose, in Google Patents (2012)
57. http://www.acelity.com/products/promogran-prisma#tab_4, in Acelity.

58. Z. Ruszczak, R. Mehrl, J. Jeckle, Multilayer collagen matrix for tissue reconstruction, in Google Patents (2003)
59. A. Dietrich, M. Myers, Modified collagen, in Google Patents (2015)
60. https://www.innocoll.com/collagen-wound-care.aspx, in Innocoll Website
61. K.V. Kavitha, S. Tiwari, V.B. Purandare, S. Khedkar, S.S. Bhosale, A.G. Unnikrishnan, Choice of wound care in diabetic foot ulcer: A practical approach. World J. Diabetes **5**, 546–556 (2014)
62. A. Reyzelman, R.T. Crews, J.C. Moore, L. Moore, J.S. Mukker, S. Offutt, A. Tallis, W.B. Turner, D. Vayser, C. Winters, Clinical effectiveness of an acellular dermal regenerative tissue matrix compared to standard wound management in healing diabetic foot ulcers: A prospective, randomised, multicentre study. Int. Wound J. **6**, 196–208 (2009)
63. S. Bottcher-Haberzeth, T. Biedermann, E. Reichmann, Tissue engineering of skin. Burns **36**, 450–460 (2010)

Chapter 5
Application of Collagen Fibril Biografts for Enhancing Local Vascularization in an *In-Vivo* Chick Chorioallantoic Membrane (CAM) Model

5.1 Introduction

Currently, one of the major problems with engineered matrices is their poor ability to become vascularized within a reasonable time [1]. There appears to be a need for biomaterials with which angiogenesis, essential for oxygen and nutrient supply, correlates with the cell invasion. The application of selected angiogenic growth factors (e.g., vascular endothelial growth factor (VEGF) and basic fibroblast growth factor (bFGF or FGF-2)) may be useful for enhancing angiogenesis [1]. However, simply admixing these growth factors to the matrices generally leads to rapid clearance from the implant site, or leaky vessels [1]. Boiling the problem down, what we need to adequately vascularize tissues is a delivery system with an ability to (1) direct the timing and release of growth factors in a controlled manner (2) and support the structure of the newly generated vascularized tissue [2].

Collagen-based biografts can deliver growth factors inducing vascularization, regulate cell behavior, as well as structurally support new vasculature and tissue regeneration [3–5]. Difficult to heal injuries, such as chronic wounds, could benefit from the design and development of a multifunctional collagen biograft that can foster rapid and functional neovascularization and tissue regeneration at the site of implantation. Neovascularization can be accelerated at the site of implantation in wounds, using growth factors (GF) which play an important regulatory role in tissue repair and regeneration in wounds (e.g., granulocyte-macrophage colony-stimulating factor (GM-CSF), platelet-derived growth factor (PDGF), vascular endothelial growth factor (VEGF), and basic fibroblast growth factor (bFGF)) [6].

Among the various growth factors, VEGF is one of the most potent proangiogenic growth factors that significantly impacts wound vascularization [7]. VEGF is a 45 kDa heterodimeric heparin-binding protein, acting as a potent mitogen (ED_{50} 2–10 pM) for micro and macrovascular endothelial cells derived from arteries, veins, and lymphatics, inducing their proliferation, migration, and tube formation

R. Joshi, *Collagen Biografts for Tunable Drug Delivery*, SpringerBriefs in Applied Sciences and Technology, https://doi.org/10.1007/978-3-030-63817-7_5

[8]. VEGF level rises in normal wound repair, leading to a vigorous angiogenic response, however, in chronic, nonhealing wounds, active VEGF level falls abnormally low, due to possible degradation of VEGF by excessively high protease activity in chronic wounds [9]. Poor vascularization which is a hallmark on chronic wounds such as diabetic foot ulcers can therefore benefit the use of long-lasting VEGF delivery [10].

The growth factors (GFs) such as VEGF are delicate to handle, costly to use, and must be administered in the correct dosages for the safety of the patient. Collectively, the safety and cost-effectiveness issues of GFs mainly derive from their initial burst release kinetics and rapid clearance at injured sites when they are administered without appropriate spatiotemporal control [11]. The short half-lives of the GFs, their relatively large size, slow tissue penetration, and their potential toxicity at high systemic doses suggest that conventional delivery techniques are not well suited to the clinical administration of GFs. To address these limitations, more sophisticated delivery systems that allow for controlled, precise, sustained, and localized release of these proteins have been developed, enabling optimal doses and spatial/temporal gradients in localized sites for effective tissue regeneration. Along with sustained delivery, the ability to tune VEGF release is also important, because wound healing varies according to the type, and age of wound, as well as many other factors including infection, sex hormones, stress, diabetes, obesity, and medications [12]. Therefore, when using collagen-based biomaterials for vascularization, it is important to control and customize the spatiotemporal release of VEGF.

VEGF can be presented in tissue, through direct immobilization in the collagen-based biomaterial. The direct immobilization approach can be achieved using three distinct strategies: physical, covalent, and bioaffinity-based immobilization (Fig. 5.1) described in detail by Chen et al. [13] and Wang et al. [11].

Of the various strategies depicted in Fig. 5.1, one noteworthy approach takes inspiration from the natural interactions between ECM and GFs (Fig. 5.1, c1). Exploiting the bioaffinity tethering between GFs and ECM via heparin binding domains, one can design optimal delivery systems capable of GF release in a highly spatiotemporal controlled manner, mimicking ECM functions.

The ECM is a highly dynamic microenvironment that regulates multiple cellular processes. It acts as a reservoir for GFs due to their ability to bind multiple molecules with high affinity. In particular, many growth factors, such as BMP-2, BMP-7, VEGF, PDG, and FGF-2, interact specifically with the heparin sulfate of the ECM. Therefore, several biomaterials have been decorated with heparin or heparin sulfate-mimetic molecules, exploiting the heparin-binding ability of GFs to improve their delivery [11].

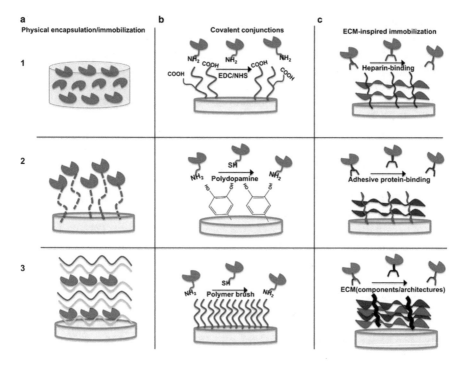

Fig. 5.1 The direct approaches for the immobilization/encapsulation of growth factors (GFs) to biomaterials. (**a**) Physical immobilization techniques. (**b**) Nonselective covalent immobilization of GFs through their functional residues. (**c**) Extracellular matrix (ECM)-inspired immobilization reactions used for the orientation of GFs on the surfaces of biomaterials. (Figure adapted from [11] licensed under a Creative Commons Attribution 4.0 International License http://creativecommons. org/licenses/by/4.0/)

5.2   The ECM, Heparin, and VEGF Regulate Vascularization In Vivo

Recently, scientists have recognized the important role of the extracellular matrix in coordinating VEGF signaling *in vivo*, including wounds [5, 14]. ECM localizes VEGF in it via heparin and heparan sulfate proteoglycan (HSPG) molecules [15, 16]. Heparin or HSPGs have a highly negative charge (approximately ≈ 75) due to the prevalence of sulfate and carboxylate groups. These groups endow heparin with an ability to electrostatically bind to many basic biomolecules, including proteins, growth factors, proteases, and chemokines [17]. The binding of VEGF to heparin occurs through such electrostatic interaction (affinity binding) [18]. Heparin then facilitates binding of VEGF to its two receptors Flt-1/VEGFR1 and Flk-1/VEGFR2 through binding and stable complex formation with neuropilin (NRP)-1 coreceptor, resulting in phosphorylation and further signaling activity of VEGF [19], such as providing essential stimulatory cues to initiate vascular branching [20] as well as endothelial tip cell filopodia emission [21]. Heparin-binding thus directly regulates the physiological effect of VEGF on endothelial cells [22–24].

Heparin also plays an important role in enabling the ECM to act as a storage depot of growth factors. Because of the affinity of heparin to type I collagen in ECM, heparin retains VEGF in ECM, protects it from proteolytic degradation [25–28], and allows the prolonged presentation of VEGF to cells [29, 30]. Heparin–type I collagen interactions likely rely on the basic triple-helical domain present at amino acid positions 87–94 near the N terminus of type I collagen monomer, and at multiple sites within native fibrils [31, 32]. Weak heparin-binding sites have been also observed near the carboxy-terminal region of monomeric tropocollagen between positions 755 and 933 [32, 33]. The regions containing elements of NH_2 terminus with an affinity for heparin are highly basic and found near the interface between the overlap and gap region of collagen (Fig. 5.2). However, these regions are also known to be participating in the cross-link formations of collagen [32]. As telopeptides and heparin are prone to bind to similar regions along the triple-helical main region, Stamov et al. proposed that heparin-binding at this position competitively inhibits the formation of asymmetric D-staggered fibrils [34]. This model is indicated in Fig. 5.2. However, competitive binding effects of heparin with collagen fibrils may change according to the concentration of heparin. While high concentrations of heparin are believed to inhibit collagen fibril assembly, low concentrations of heparin have been reported to be promoting fibril formation [35–38].

5.3   VEGF Isoform Selection for Vascularization

VEGF (vascular endothelial growth factor)-A is the most prominent member of the PDGF (platelet-derived growth factor)/VEGF family of secreted dimeric growth factors that supports vascularization. VEGF exists as multiple biochemically distinct protein isoforms that differ in their ability to interact with NRPs and HSPGs [40].

The VEGF isoforms produced are a result of alternative mRNA splicing of a common transcript from a single *VEGF* gene. In humans, although multiple spliced isoforms have been identified, the most common and well-studied isoforms are composed of 121, 165, and 189 amino acids. The eight exons of the human VEGF-A are alternatively spliced to produce several distinct isoforms that differ in the presence or absence of exon 6 and 7, the Heparin-Binding Domains (HBDs). Longer VEGFA splice isoforms (VEGFA165 and VEGFA189) contain heparin-binding domains that allow for ECM binding and domains for binding to the co-receptor neuropilin 1(NRP-1) [40, 42], as illustrated in Fig. 5.3.

VEGF 165 binds to heparin through positively charged lysine and arginine residues encoded by exon 7 of the VEGF gene [43]. However, it is also known that VEGF 189 isoform contains in addition to the amino acids encoded by exon 7, 24 amino-acids that are derived from exon 6, constituting yet another heparin-binding domain [44]. As a result, VEGF 189 shows a stronger affinity for heparin due to the presence of two heparin-binding domains [45, 46], and these binding sites are

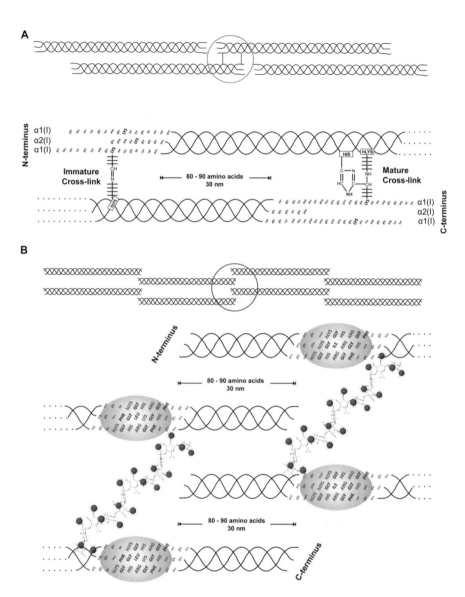

Fig. 5.2 Molecular origins for the native tropocollagen and in vitro atelocollagen-heparin fibrillar assembly. (a) The classical staggering of native collagen fibrils involves later stabilization via an immature aldimine cross-link between Lys-aldehydes from the telopeptides and the triple-helical hLys. The bond is stable at physiological conditions, although it can further "mature" by involving a triple-helical His to form histidinohydroxylysinonorleucine (HHL). The bond is more characteristic of the C-terminus but can also be found in the N-terminal region. (b) A schematic representation of the proposed heparin-binding clusters, which appear to be almost symmetrically allocated at both the C- and the N-termini. The absence of telopeptides creates a highly basic binding pocket that could be a favorable site for binding of the highly negatively charged heparin molecule. In both (a) collagen and (b) collagen-heparin, there is a triple-helical terminal overlap of 80–90 amino acid residues. The skeletal model of heparin in (b) was constructed using previously reported NMR coordinates [39], emphasizing a Van der Waals surface representation of the negatively charged sulfate groups. The collagen monomers and heparin are not drawn to scale so that the available functional groups and amino acid residues can be visualized more clearly. (Figure adapted with permission from Stamov et al. [34])

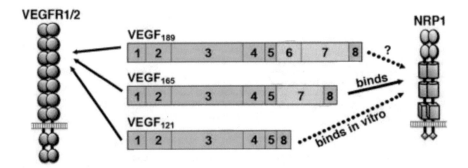

Fig. 5.3 Current knowledge of VEGF isoform binding to their receptors. All isoforms bind VEGFR1/2, whereas only VEGF165 and possibly VEGF189 is known to bind NRP1. VEGF121 can bind NRP1 with low affinity in vitro, but whether this association occurs in vivo has not been shown. (Figure adapted from Tilo et al.'s work [41], an Open Access article distributed under the terms of the Creative Commons Attribution License (http://creativecommons.org/licenses/by/3.0))

reported to be distinct from VEGF's receptor-binding domain [47]. In contrast, the shorter isoform VEGF121 is not yet shown to bind to the ECM *in vivo*.

The heparin-mediated localization of VEGF within ECM results in concentration gradients. VEGF165 and VEGF189 form steep concentration gradients within the extracellular space, remaining close to the site of production (e.g., VEGF189 accumulates in the basement membrane), whereas VEGF121 forms shallow concentration gradients [24]. Differences in VEGF localization are thought to give rise to vascular phenotypes that vary with the VEGF isoform in a monotonic fashion. Larger isoforms, with greater heparin-binding affinity, give rise to sprouts that have greater filopodial directionality and vessels with greater branching density and smaller diameters [23] as illustrated in Fig. 5.4.

Inspired by the coordinated role of ECM, heparin, and VEGF in providing biochemical and biomechanical cues for in vivo vascularization [48–51], many scientists have prepared collagen-based materials with heparin incorporated in them for loading of VEGF [1, 52–57]. However, these systems consist of monomeric collagen formulations that are chemically cross-linked for the retention of heparin. As a result of chemical cross-linking, VEGF had to be loaded in the last step of formulation, so that chemicals used for cross-linking would not destroy the bioactivity of VEGF. VEGF is then loaded typically either through immersion of the formulated matrices into VEGF solution or through impregnation of VEGF into the collagen matrices, both of which can lead to low VEGF loading efficiency. These strategies and their limitation are depicted in Table 5.1 and discussed in section 5.4.

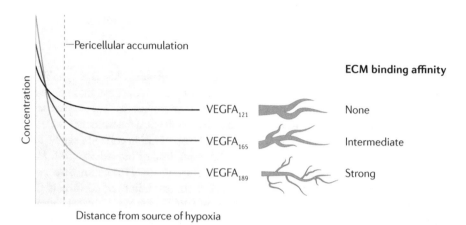

Fig. 5.4 Changes in the expression of vascular endothelial growth factor A (VEGFA) splice isoforms with a varying affinity for the extracellular matrix (ECM) alter growth factor concentration gradients within tissues and the resulting vascular morphology. VEGFA121 does not bind to the ECM, creating shallow VEGFA gradients in tissues and resulting in the formation of wide-diameter vessels with low branching density. Conversely, VEGFA189 binds strongly to the ECM, resulting in short, steep VEGFA gradients and the formation of networks with thick, highly branched vessels. (Figure adapted with permission from Briquez et al. [24])

Table 5.1 Selective strategies used for VEGF delivery from collagen-based delivery systems

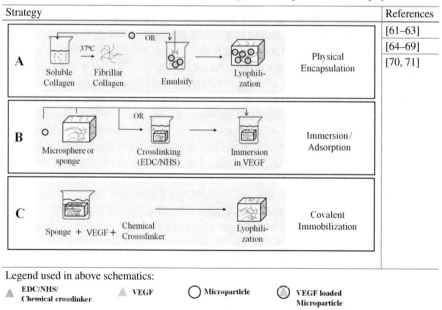

5.4   Improving VEGF Retention Capacity of Collagen Implants

For improving the vascularization ability of collagen scaffolds through VEGF incorporation, a variety of approaches have been adopted in the past, including simple physical entrapment, adsorption, and covalent immobilization. These approaches are summarized in Table 5.1. While being simple, the physical entrapment (Table 5.1, approach A) or adsorption (Table 5.1, approach B) of VEGF in collagen can be ineffective due to its rapid outward diffusion and quick loss of bioactivity [58]. More serious problems such as abnormal, tortuous, and leaky vessel formation on account of the uncontrolled release of VEGF can lead to clinical failure of constructs [24]. Therefore, to prevent the uncontrolled release of VEGF, covalent immobilization of VEGF has been developed (Table 5.1, approach C). While chemical immobilization can prevent passive diffusion of VEGF, it comes with a disadvantage that chemical cross-linkers alter the inherent biological signaling capacity of collagen and can result in adverse tissue responses [59, 60]. Moreover, it also presents a danger of damaging the functional group or the screening of the active pocket of the VEGF.

To surpass these limitations, heparin-based retention of VEGF in collagen implants has emerged as an attractive option recently. These approaches are represented in Table 5.2. It should be noted that for heparin incorporation in collagen, several studies (Table 5.2A–D) have used a chemical cross-linker called 1-Ethyl-3-(3-dimethylaminopropyl) carbodiimide (EDC) in combination with N-hydroxysuccinimide (NHS) that activates heparin for immobilization [1, 52–57]. While this cross-linking also serves to improve the mechanical strength and proteolytic resistance of conventional collagen formulations, it alters the native physiological structure of collagen due to chemical cross-linkage [72]. As a result, fibrillar mechanics are also affected, and since cell traction forces and adhesive behavior depend on these fibril mechanics, any alteration to this native structure of collagen can also affect cell proliferation and movement [34]. Furthermore, due to EDC/NHS chemical cross-linking step, to preserve biological activity, the VEGF has to be loaded to the collagen matrices in the last step, through immersion or impregnation, resulting in a lower loading efficiency. Moreover, it is currently not possible to use EDC chemistry to independently vary the implant stiffness versus the amount of immobilized VEGF [68]. Finally, the heparin quantities used for VEGF retention are high, and the effect of heparin on collagen fibril mechanics is not always given, although it is now known that heparin can alter both the microstructure and mechanical properties of collagen [34, 36–38, 73, 74].

These issues can be addressed through a design strategy purely relying on affinity-based retention of heparin and VEGF as opposed to using exogenous chemical cross-linkers. Exploiting heparin affinity for collagen and VEGF, heparinized collagen implants containing VEGF can be created (Table 5.2, Strategy E) that retain their physiologically relevant self-assembly properties.

Table 5.2 Selective strategies used for VEGF delivery from heparinized collagen materials

Strategy	VEGF loading	References	Limitation
A		[75, 76]	Collagen self-assembly not capitalized;
B	Immersion or Impregnation	[77, 78]	Chemical cross-linking used;
C	(Affinity-based)	[1, 54–58]	High-heparin quantity;
D		[53, 79]	Low-VEGF loading efficiency;
E	Affinity-based	[80]	Slow vascularization

Legends used in the above schematics:

SC = Soluble Collagen   EDC/NHS/   ▲ VEGF   Oligomer with
FC = Fibrillar Collagen   Chemical crosslinker   natural crosslink

It should be noted that many of the systems mentioned in Tables 5.1 and 5.2 have performed in vitro tests before going in vivo. When evaluating a new collagen-based drug delivery system for vascularization, we recommend doing animal tests with ethical considerations. While the studies performed in vitro can generate enormous data when testing a molecular delivery system, the true therapeutic potential of a system can be analyzed via in vivo experimentation which offers a stronger, reliable value. Starting with a simple model such as the chicken egg Chorioallantoic membrane (CAM) model is recommended, since the CAM model has been used effectively by several disciplines including biology, medicine, and bioengineering, offering an exceptionally valuable in vivo testing tool.

5.5 Chicken Chorioallantoic Membrane (CAM) Vascularization Assay

The CAM assay offers many advantages over other *in vivo* models used to study angiogenesis and vascularization. The most important advantage of the model is its accessibility and rapid growth. The CAM develops in a short time (a period of only 7 days or Embryonic Day 3–10), during which CAM develops from a small avascular membrane into a structure that covers the entire inner surface of the shell displaying a densely organized vascular network. This enormous angiogenic boost provides a wealth of information related to the required developmental pathways. CAM model can be used to study pathological processes upon simple exposure to cytokines, hormones or drugs, or transplantation of tissues, isolated cells, or materials [81].

The main function of the CAM is to serve as the respiratory organ for the chicken embryo. The CAM also helps in the storage of excretions, electrolyte transport (sodium and chloride) from the allantoic sac, and mobilization of calcium from the shell to start bone mineralization. It is connected to the embryonic circulation by the allantoic arteries and veins associated with lymphatic vessels [82]. During embryo development, many features of the CAM change, such as the composition of the ECM, the degree of differentiation of endothelial cells and vessels, the characteristics of the inter-endothelial junctions, as well as the location of the vessels within the CAM [82]. This issue emphasizes the importance of always using embryos at the same Embryonic Development Day (EDD), failing which, comparisons between experiments can become problematic.

The chick CAM is a part of the extra embryonic tissue that begins to develop 7 days after initial incubation from the fusion of the chorion and the allantois. Structurally, the outer epithelial layer of the chorion is derived from the trophoblast, which opposes the allantois. This structure forms a supportive matrix for the extensive vascular network that courses through the CAM (Fig. 5.5 inset). The mature chick CAM (20–100 μm) and human retina (approximately 100–300 μm) are of roughly comparable thickness [83]. Mature chick CAM (incubation day 12 and on) can be divided into three anatomically distinct layers: (1) primary stratum; (2) capillary plexus, or blood sinus; and (3) a thin stratum composed primarily of specialized chorionic epithelial cells that have presumably migrated above the capillary plexus/

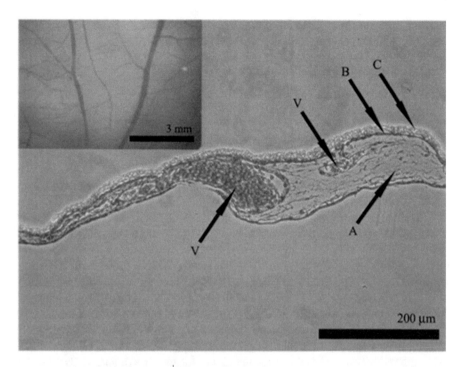

Fig. 5.5 Chick chorioallantoic membrane (CAM). Photograph of mature chick CAM (inset) demonstrating an arborizing network of vessels. Histologic cross-section of mature chick CAM demonstrating large and small blood vessels (V), primary stratum (A), thin stratum (B), and inner shell membrane (ISM) (C). A capillary plexus/blood sinus exists between the primary (A) and thin (B) layers. Some erythrocytes in this plexus/sinus are faintly visible as circular bulges around the thin stratum (B). (Figure adapted with permission from Leng et al. [83])

blood sinus and are involved in gas exchange and calcium absorption [83]. Immediately above and attached to the CAM thin layer is the inner shell membrane (ISM) (Fig. 5.5).

The CAM model offers ease handling for both intervention and imaging of the vasculature. In particular, the broad applicability of distinct imaging modalities ranging from microscopic – to magnetic resonance – to positron emission tomography imaging (MRI and PET) have attracted many researchers. Besides, due to the transparency of its superficial layers, nearly any wavelength in the visible part of the electromagnetic spectrum can be used for fluorescence imaging. For applications related to transplantation, an important benefit is the slow developmental progression of the immune system that reaches physiological activity only by day 15 post-fertilization [84]. Thus, unwanted immunological reactions are limited, facilitating the transplantation and study without compromise. Last but not least, the cost-effectiveness of the model adds to the long list of benefits. The price of a fertilized egg is 100 times lower than that of a mouse of a common strain.

However, the CAM model also has limitations. From the standpoint of studying the effect of a drug released on CAM, the already well-developed vascular network of CAM can make it difficult to differentiate between new capillaries and already existing ones. Another limitation is that if experiments extend after 15 days, a non-specific inflammatory reaction can occur that may limit the success of grafting. Such reactions may provoke angiogenic responses, which are difficult to distinguish from an angiogenic activity of grafted material [85].

Nonetheless, considering the enormous benefits offered by the CAM model, the molecular delivery of VEGF from collagen matrices and their ability to induce vascularization in vivo can be studied by setting up an experiment in a CAM model. Controls such as paper disc and commercial collagen sponges can be added to the experiment. The CAM assay is briefly mentioned below, based on the procedures described elsewhere [86, 87].

Fertilized White Leghorn chicken eggs are horizontally positioned and incubated at 38 °C under 58% ± 2% relative humidity in an egg incubator equipped with a turner which automatically rotates the eggs 5 times/day until day 7 [88]. On day 8, a window of approximately 2.5 cm diameter is created using a Dremel tool equipped with a cutting disc. The window is then sealed with sterile adhesive tape to prevent loss of moisture and eggs are returned to the incubator. This allows us to see if any adverse reaction is generated on CAM and viability is compromised due to eggshell dust fallen during cutting of the air sac. On an embryonic day 9, collagen implants are inserted on the CAM of viable eggs. Post 3 days of implantation (on embryonic day 12), all CAM samples are photographed and fixed in situ using 4% paraformaldehyde [89]. Each experimental group can be assigned 10–12 eggs to start with, with approximately 6–8 viable eggs remaining for the experiment on day 12. An example timeline of CAM assay is shown in Fig. 5.6 part A, while part B shows the localization of CAM in the chicken egg during various stages of the assay [81].

5.5.1   Care During CAM Experimentation

Some nuances of CAM Assay should be mentioned for enhancing the efficacy of CAM assay. Fertilized chick eggs should be kept in adequate humidity, a 37 °C environment, and rotated every few hours, which can be performed in an inexpensive incubator. During *in ovo* incubation, the eggs should be slowly rotated until the day of cutting the air sac and implanting biografts on CAM. Turning of the eggs prevents the embryo from sticking to the shell membranes which would occur if it is left in one position too long. For good results, it is recommended to turn the eggs the first thing in the morning, again at noon, and the last thing at night [85]. Alternatively, the use of an automatic rotator enables the slow but constant movement of the eggs.

To prepare CAM for a typical experiment, the eggshell is cracked and peeled away from the region over the air space that exists between the shell and the inner shell membrane (ISM) at one pole of the egg. This airspace can be visualized before the egg is cracked by holding the egg under intense light, for example, surgical light, or handheld flashlight. This process is sometimes referred to as candling of the eggs and we recommend marking the air sac with a pencil or a marker while candling the

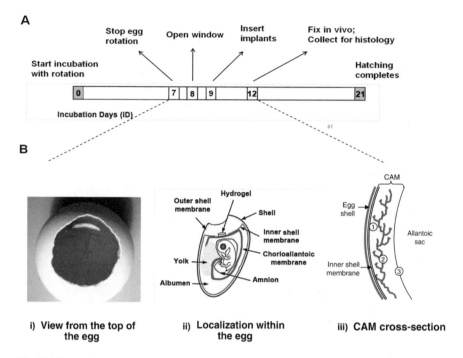

Fig. 5.6 The schematics of the CAM assay timeline. (a) Timeline involves incubating eggs until Embryonic Day (ED) 7, Candling the eggs to mark the air sac, then cutting a window open on ED 8, inserting collagen implants on ED 9, returning the eggs to the incubator, and finally evaluating the CAM on ED 12 after fixing in situ. The CAM localization between the assay steps for the embryonic day (ED) 8–12 is shown in part B. Part B (i) shows a CAM appearance when a window is cut open in the air sac of the egg, and Inner Shell Membrane is peeled on ED8. Part B (ii) shows the implantation of collagen implant on CAM on ED9. Part B (iii) illustrates a cross-section of the CAM at day 12 of incubation. 1: chorionic epithelium; 2: mesoderm with blood vessels depicted in red; 3: allantoic epithelium. (Figure B parts (i) and (iii) reproduced with permission from Vargas et al. [81], and part B (ii) reproduced from [90] with permission from the Royal Society of Chemistry)

egg. The marking will make it easier to cut into the air sac of an egg without damaging the CAM.

Once a window is cut into the air sac, an opaque inner shell membrane (ISM) is exposed. This inner shell membrane (ISM) is positioned immediately above the CAM and is attached to the CAM (Fig. 5.7). For accessing the CAM below ISM, the opaque CAM-ISM dual-layer should be first irrigated with saline. The saline irrigation will cause the dual-layer to become translucent, allowing for visualization of the CAM vasculature [83]. Depending on the requirements of the proposed experiment, the ISM can be left in place or peeled away from the CAM with the use of fine forceps as shown in Fig. 5.7 below. Since there is an extensive blood sinus/capillary bed between the CAM primary stratum layer and the ISM, peeling the ISM may cause temporary minor bleeding in some eggs. This bleeding, however, usually

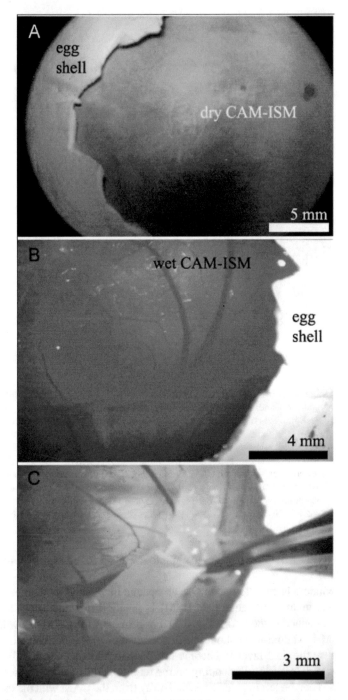

Fig. 5.7 Preparation of CAM. (**a**) To prepare CAM for a typical experiment, the eggshell is cracked and peeled away from the region over the air space that exists between the shell and the ISM near one pole of the egg. (**b**) Irrigation of the CAM-ISM with saline causes the dual-layer to become translucent, allowing for visualization of the CAM vasculature. Experiments can be performed at this stage. (**c**) The ISM can also be peeled away from the CAM with the use of fine forceps. (Figure adapted with permission from Leng et al. [83])

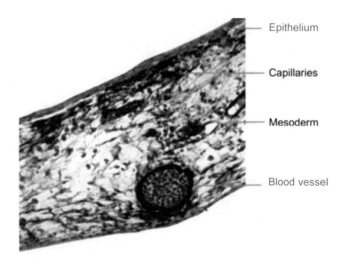

Fig. 5.8 Immunohistochemical visualization of α-smooth muscle actin (*dark brown*) in a transversal CAM section. Smaller capillary plexus is more superficial in comparison to the larger vessels located deeper in the CAM. (Figure adapted with permission from Norwak-Sliwinska et al. [85])

stops within 1–3 min. Introducing fresh saline into the air space will wash away the blood, providing a clean membrane for experimentation [83].

Care should be also taken to maintain 55–60% humidity in the egg-incubator. Proper humidity eliminates extra barriers in the delivery of soluble molecules to the underlying vascular tree of CAM, as explained histologically. The CAM histology shows two epithelial sheets that limit a thin layer of the stroma. The upper epithelium is of ectodermal origin, while the stroma and the lower epithelium are of mesodermal and endodermal origin respectively (Fig. 5.8). It is within the stroma that the blood vasculature and lymphatics reside [85]. Therefore, it is important to realize that any compound delivered on the surface of the CAM has to pass across the surface epithelium and reach the vessels in the stroma. Poor humidity conditions when the CAMs are cultured *ex ovo* induce significant cell division and keratinization of this upper, and normally single, epithelial layer; making it even more difficult to deliver soluble molecules to the underlying vascular tree [85]. Therefore, the humidity status is critical during the experimental use of this system.

There are also alternative protocols to *in ovo* CAM cultivation assay. Ultimately, the goal and type of experiment direct the choice of the protocol. In an alternative protocol, CAM can be cultured *ex ovo* where investigators rupture the egg and transfer the embryo to a 10 cm petri dish at day 3 post-fertilization (Fig. 5.9). This *ex ovo* culture of the embryo (also known as shell-less embryo culture, Fig. 5.9a–c) offers ample observation and manipulation space, allowing for testing of multiple implants/samples in a single CAM, compared to *in ovo* CAM protocol (Fig. 5.9d–f). However, removing the embryo from the shell is associated with a low survival rate, due to the frequent rupture of the yolk membrane either during or after culture [85]. This occurs due to excessive distension of the yolk in the flat dish. *Ex ovo* survival rates

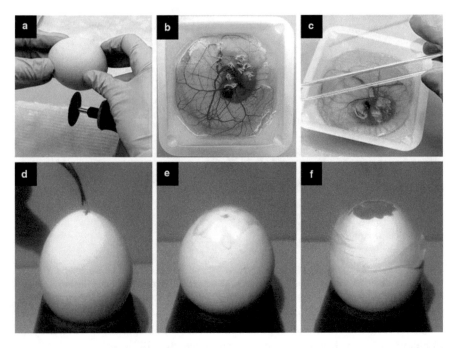

Fig. 5.9 CAM cultivation protocols. The *ex ovo* protocol (**a–c**) requires (**a**) a gentle cracking of the egg using a blade and transferring of the contents into a petri dish or weighing boats (**b**) with a plastic cover (**c**). The *in ovo* cultivation (**d–f**) starts with gentle rotation for 3 days. On embryo development day 3, the eggshell is opened with sterile tweezers (**d**) and subsequently covered with a laboratory wrapping film (**e**). On the day when the experiment starts, the shell above the air pouch of the egg is extended to a diameter of approximately 3 cm, (**f**) enabling further experimental manipulation. (Figure adapted with permission from Norwak-Sliwinska et al. [85])

can be drastically improved through reducing the impact during the cracking of the eggshell and placing the embryo in a confined curved space rather than a flat petri dish for culture. Gentle cracking of the egg can be achieved using a circular saw rather than cracking the egg against a sharp edge (Fig. 5.9a). Alternative cultivating dishes include large weight boats (Fig. 5.9b) and plastic inserts that can be placed inside 100 mm petri dishes to introduce curvature to the dish (Fig. 5.9c). Besides, it is of utmost importance to keep the humidity in the incubator at 98% or higher and to maintain the culture environment sterile. Using these approaches the viability is around 50% on day 14. For a demonstration of a refined chick *ex ovo* culture and CAM assay with survival rates over 50% refer to the video article of Dohle et al. [91].

In ovo cultivation is an important alternative that improves significantly the survival of the embryos (Fig. 5.9d–f). In an alternative *in ovo* protocol, at day 3 post-fertilization, a hole of approximately 3 mm in diameter is created in the eggshell with sterile tweezers and covered with a laboratory wrapping film to prevent dehydration and possible infections (Fig. 5.9d). This initial small incision is said to change the pressure inside the egg and prevent the binding of the CAM with the shell membrane. The eggs are then returned to the incubator with a relative air

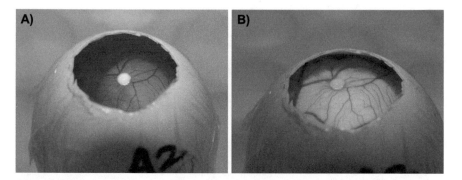

Fig. 5.10 Heavy cream enhances contrast for photographing CAM. The CAM was implanted with an oligomer collagen construct on Embryonic Day 9, returned to the incubator, and after 3 days, photographed on Embryonic Day 12 without heavy cream (**a**). Underneath the same CAM was injected a warm heavy cream: distilled water in a 50: 50 ratio at 37 °C, which enhanced the contrast to capture the view of the blood vessels of CAM around the collagen implant (**b**). (Figure based on own unpublished work)

humidity of 65% and a temperature of 37 °C in a static position until use (Fig. 5.9e). At day 7 or later, the hole is extended to a diameter of approximately 3 cm to provide access to the chorioallantoic membrane (Fig. 5.9f). This approach enables experimentation of an implant through a window in an almost unchanged physiological environment for the developing embryo.

After the completion of the implantation period, photographic visualization of CAM can be obtained by injecting emulsified fat (30–50% heavy whipping cream diluted in distilled H_2O) warmed to 37 °C right under the CAM. This tactic provides enhanced contrast for viewing CAM vasculature by giving a white background underneath the CAM, and in part obstructing the underlying vessels that are not associated with the experiment [92, 93]. The difference between a CAM photographed with and without heavy cream for contrast is shown in Fig. 5.10.

The type of CAM experiments described herein can optimally be performed before incubation day 18. After day 18, the embryo is large enough that its size and movements underneath the CAM can disrupt experimental maneuvers on the CAM. Additionally, standard animal protocols for chick embryos past day 18 require more complicated euthanasia techniques (chicks typically hatch between days 20 and 22).

5.6   CAM Vascular Response to Collagen

The vascularization potential of collagen biografts can be studied using the assay described above, revealing if a sample elicits a spoke-wheel pattern response of CAM vasculature around the implant. For example, Joshi applied low- and high-fibril density collagen implants containing heparin and VEGF on CAM at ED 9 and

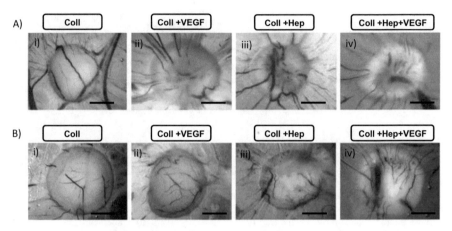

Fig. 5.11 Heparinized, VEGF containing low- and high-fibril density collagen implants show improved vascular response on CAM without inducing abnormal vessel formation. CAM assay procedures were performed as described elsewhere [86, 87]. CAM was implanted with 3 mg/ml oligomer collagen implants (**A**) and 20 mg/ml oligomer collagen implants (**B**), incubated for 3 days, then fixed in situ with 4% paraformaldehyde and viewed under a stereomicroscope. Implant groups in **A** and **B** consist of oligomer (i), oligomer +0.5 µg/ml VEGF (ii), oligomer +1 µg/ml heparin (iii), and oligomer +1 µg/ml heparin +0.5 µg/ml VEGF (iv). Implants were fixed in situ and harvested CAM was washed in saline and imaged using a stereomicroscopic camera. Tortuous vessels were observed in VEGF-loaded implants without heparin (**A**-ii, and **B**-ii), while VEGF-loaded implants with heparin showed a clear spoke wheel pattern of vascular response (**A**-iv, and **B**-iv). Scale bar represents 1.25 mm in panel **A** and 2.5 mm in panel **B**. (Figure based on unpublished own work)

studied the CAM vasculature response to the implant after 3 days of exposure [80]. It was observed that collagen implants with Heparin and VEGF together in them (Coll + Hep + VEGF group) elicited a strong vascular response in the form of a spoke wheel pattern (Fig. 5.11 A-iv and B-iv). This could be attributed to the successful retention of VEGF in the implant, and its subsequent sustained, slower release on CAM. Collagen implants with heparin alone (Coll + Hep group) showed a weaker vascularization response (Fig. 5.11 A-iii and B-iii), with thinner vessels drawn toward the construct. The oligomer alone (Coll group) showed the lowest spoke-wheel pattern form of a response (Fig. 5.11 A-i and B-i), while oligomer with VEGF (Coll + VEGF) showed vascular response which was found to be abnormal due to tortuous, or irregular appearance of the vessels (Fig. 5.11 A-ii and B-ii).

For the clinical success of tissue-engineered scaffolds, along with the promotion of vascularization around the implant, the growth of microvessels within the implant is crucial to enable the survival of cells in the core of the scaffold [24]. CAM assay allows the advantage of envisioning such a microvasculature growth inside scaffolds that can be separated from the surrounding CAM vasculature, as these microvessels grow inside the scaffolds against gravity [94–96]. The vascularization response can be determined using two scoring schemes to compare photographs from day 12 taken post-implantation of collagen matrices in CAM, to those from

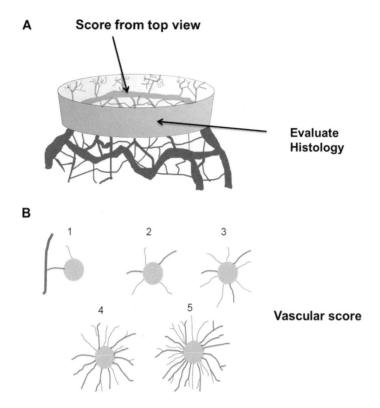

Fig. 5.12 Implant evaluation for a vascular response. (a) Implants can be evaluated for their vascularization ability by scoring CAM vessel response from the top view, before performing histological staining. (Figure **a** adapted with permission from Norwak-Sliwinska et al. [85]) (**b**) Drawings representing examples of different vascular responses in CAM assay. A ranking method from 1 to 5 was used for semiquantitative scoring of vessel density and distribution of CAM around the implanted inserts

day 9. A "vascular score" is determined by scoring the vessel density and distribution (5 = strong, 1 = weak) around the collagen implants on CAM [94], by observing the top view as shown in Fig. 5.12. Similarly, vessel tortuosity and abnormality in CAM can be scored on a scale of 1–5 (5 = irregular/tortuous/brush-like, 1 = normal), based on observation of irregular vessels or fine brush-like vessels on CAM.

Several other visual vessel-counting methods or automated approaches can be used for CAM assay. These are summarized in various reviews [85, 94, 97]. Quantification has frequently relied on descriptors such as vascular density, vessel branching points/mm^2, vascular length, or endpoint density/mm^2, please refer to [85] for details. One example is presented in Fig. 5.13. Here, Kim et al. [52] formulated a composite collagen scaffold for delivering angiogenic factors in CAM. Vascular endothelial growth factor (VEGF) was loaded on mesoporous silica nanoparticle (MSN), which was then incorporated within a type I collagen sponge, to produce collagen/MSN/VEGF (CMV) scaffold. The CMV composite

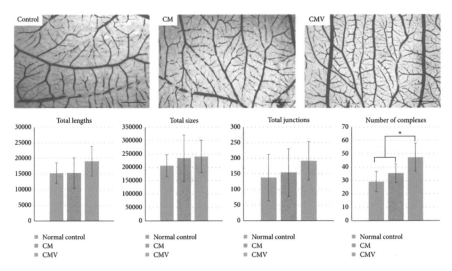

Fig. 5.13 Representative images of blood vessel networks formed in the CAM model after 4 days of sample placement in contact with embryos (×10, bar = 200 μm). Image analysis represents the total length, total size, total junction, and the number of complexes of blood vessels for control, CM, and CMV group (mean ± standard deviation, compared to CMV). (Figure adapted from work by Joong-Hyun Kim et al. [52] covered under the Creative Commons Attribution License)

scaffold was observed to release VEGF sustainably over the test period of 28 days *in vitro*. The release of VEGF improved cell proliferation. Moreover, the in vivo angiogenesis of the scaffold, as studied by the chick chorioallantoic membrane (CAM) model, showed that the VEGF-releasing scaffold induced a significantly increased number of blood vessel complexes when compared with the VEGF-free scaffold. These results demonstrate that the collagen scaffolds with VEGF-releasing capacity can be potentially used to stimulate angiogenesis and tissue repair.

5.7   Heparinizing Collagen Grafts to Promote Vascularization in CAM

The positive angiogenic effect of heparin on CAM vasculature is known [98–100]. In vivo, heparin-based VEGF retention has been found to increase endothelial cell proliferation, upregulate microvasculature formation, and stimulate blood vessel maturation [53, 101–103]. Even in the absence of exogenous growth factors, modification of collagen with heparin was found to increase neovascularization, possibly by potentiating endogenous growth factors present in vivo [3, 53, 54]. This positive effect of heparin could have been due to its role in protecting cell-secreted VEGF from degradation [58], and upregulating VEGF activities by enabling its binding to the KDR and Flt-1 receptors [104]. These facts point at the potential of heparin in retaining VEGF in collagen implants and upregulating its activities on CAM.

In general, the vascularization potential of heparinized collagen implants can be beneficial in cases such as small injuries or as acute wounds, where the low mechanical properties of the implants could suffice tissue healing for a short period. However, in cases such as chronic wounds, where the regeneration of new tissue is difficult due to a high level of proteases [105, 106], the implants would be required to last longer to support and accelerate new capillary ingrowth into the implant. For this purpose, collagen implants of high fibril density can offer a potential solution due to their characteristic higher mechanical strength and resistance to proteolytic degradation [107]. Moreover, the increased fibril-density would have a positive effect on retaining the encapsulated growth factors [108], due to their enhanced fibril density (reduction in pore size) [109].

5.8   A Note on Selecting Effective VEGF and Heparin Quantity

VEGF dose should be carefully chosen in CAM study because of the evidence that overdosage of VEGF therapy can result in an imbalance in angiogenic signals, leading to dysregulated vasculogenesis [110, 111] and hemangioma-like assemblies [110, 112]. Rapid, unregulated exposure of freely diffusible VEGF has been previously reported to cause excessive but abnormal, unstable blood vessel growth [113, 114]. It is therefore helpful to consult a various range of concentrations of VEGF reported in assays that showed enhanced vascularization effects in vivo either with heparin [1, 52–57] or without heparin [65–71, 115]. In vivo, heparin based VEGF retention has been found to increase endothelial cell proliferation, upregulate microvasculature formation, and stimulate blood vessel maturation [53, 101–103]. However, in vitro addition of heparin has illustrated that these effects are concentration-dependent and beneficial effects were found only at low concentrations (0.1–1 µg/ml) of heparin [18, 116], while higher concentrations (10–1000 µg/ml) of heparin progressively inhibited the VEGF binding [19, 117]. These results prompt one to carefully select heparin concentration for admixing to collagen, to obtain beneficial effects of VEGF binding.

Another important consideration in selecting heparin concentration for addition to oligomer is its effect on collagen fibril self-assembly. Heparin is known to bind to type I collagen fibrils with high affinity (Kd = 150 nM) [31]. However, several investigators over the past few decades have reported that the presence of heparin during collagen fibrillogenesis in vitro could have profound concentration-dependent effects on fibril size, interconnectivity, diameter, and organization [34, 36–38, 73, 74], that could impact cell growth [37]. Stamov et al. recently reported that these gross physiochemical and morphology changes could be attributed to the competitive binding of telopeptides and heparin to similar regions along the triple-helical main region of intact tropocollagen, leading to inhibition of the formation of asymmetric D-staggered fibrils [34].

Heparin bears the highest density of negatively charged groups among all other GAGs [118] which can trigger the electrostatic interaction with other macroionic molecules such as collagen [119]. The profound effects of heparin on the processes of collagen fibril formation, growth, and higher-level organizations of collagen matrices were found to be concentration-dependent [34, 36–38, 73, 74]. While low concentrations of heparin were reported to be promoting fibril formation, high concentrations inhibited fibril assembly [35–38]. However, the concentration range of heparin, the ratio of collagen and heparin, as well as the investigation techniques varied considerably in these studies, making it difficult to paint a consistent picture of the important parameters of heparin interaction with collagen. Therefore, before employing a strategy of heparin-based VEGF retention in collagen, it is extremely important to find the effect of heparin on that collagen formulation (self-assembly/polymerization) and viscoelastic properties.

While it is known that heparin and collagen form stable complexes due to electrostatic interactions between the highly anionic heparin and the positively charged groups of collagen [120, 121], the possibility that not all the added heparin binds to oligomer –should be taken into consideration. Some heparin could be merely physically entrapped in the polymerizing matrix. Since this free (unbound) heparin in the collagen matrix could result in its uncontrolled diffusion out of collagen, it is necessary to eliminate any unbound heparin from the collagen matrix using a large excess of washing solution, such as 1X phosphate buffer saline. To quantify the remaining heparin in washed matrices, a DMMB assay can be performed exploiting the fact that heparin forms colored complexes with the cationic dye 1, 9-dimethylmethylene blue. All samples need to be papain digested before the assay to make the entire amount of heparin present in the matrix accessible for the dye complexation [122].

5.9   Cellularization and Vascularization Within the Collagen Constructs

For the clinical success of tissue-engineered scaffolds, the growth of microvessels *within* the collagen implant (not merely around the implant), is crucial to enable the survival of cells in the core of the scaffold [24]. CAM assay allows the advantage of envisioning such a microvasculature growth inside scaffolds that can be separated from the surrounding CAM vasculature, as these new microvessels grow inside the scaffolds against gravity [94–96].

While several studies reported that modified collagen implants increased in neovascularization on the CAM surface [53], very few have documented actual neovascularization inside the collagen matrix of the implant [123]. Kilarski et al. [123] reported that neovessels found in their collagen matrix were contained within the expanding CAM tissue that eventually replaced the provisional matrix and there was a clear demarcation between the ingrowing tissue and the implanted gel, implying that the neovessels entered their collagen gel as a part of ingrowing CAM tissue buds, not as independent entities.

Kilarski et al. [123] used a wound-healing model based on the chick chorioal-lantoic membrane (CAM) and investigated early steps of nondevelopmental tissue neovascularization (Fig. 5.14). First, they formed a provisional matrix composed of fibrinogen (5 mg/ml) and rat tail collagen I (4 mg/ml) in a plastic tube glued on a nylon mesh; this assembly was then placed on the CAM at 12 d of embryo develop-ment through a window in the shell of the egg (Fig. 5.14a). The nylon mesh sepa-rated the CAM from the provisional matrix, which allowed discrimination between new and preexisting vessels. After the addition of FGF-2 to the matrix, eggs were incubated for 6 d. Whole eggs were then fixed, the matrix constructs were cut out from the CAM and the plastic tube was removed. In response to fibroblast growth factor-2 (FGF-2), the gel gradually contracted and was partly digested by invading cells (Fig. 5.14b). In parallel to these physical changes in the gel, they observed ingrowth of vascularized tissue evidenced by the visualization of blood-filled struc-tures through permanent staining of erythrocytes with 3,3'-diaminobenzidine (DAB) and H_2O (Fig. 5.14c–g, i, j).

Tissue contraction depends on both the imposed tension and compliance of the matrix to deformation. Kilarski et al. observed that gel contraction was a prerequi-site for neovascularization [123]. They observed lower cell-driven contractibility of Vitrogen due to pepsin-mediated cleavage, which results in the removal of both collagen telopeptides and cryptic binding sites for integrins. This loss of integrin binding sites, and collagen telopeptides hinders the cells to translocate matrix mol-ecules, thus showing lower cell-mediated tension in the matrix. Kilarski et al. also reported the impaired contractibility of Vitrogen *in vitro* using gels containing chicken-derived myofibroblasts (Fig. 5.15b). They observed a lack of contraction in the Vitrogen-based gel paralleled by a failure of cells to adopt a concentric orienta-tion parallel to the gel surface, which they mentioned necessary to reduce the sur-face area by contraction (Fig. 5.15c). Myofibroblast invasion or migration in vivo was not inhibited in Vitrogen as indicated by their αSMA-positive cells found in the Vitrogen gel. These cells probably acquired their myofibroblastic phenotype as they migrated and differentiated in the activated environment of the underlying CAM but failed to spread when they entered the implanted matrix and instead appeared as rounded cells (Fig. 5.15d).

Similar to the cellularization of FGF-loaded collagen constructs, VEGF-loaded collagen constructs can also trigger cellularization within the implants. Joshi [80] observed that a cellularization of collagen constructs, when placed on CAM, can result in the contraction of the implant due to CAM cell infiltration as suggested by Kilarski et al. [123]. Cellurization was highest in the Coll + Hep + VEGF group (Fig. 5.16d) of 20 mg/ml implants which also exhibited the highest % contraction (66 ± 9.16%) among the high-fibril density group. The Coll + Hep group showed the second-highest contraction that also correlated with this group's second highest cel-lular infiltration (Fig. 5.16b). The contraction and cellular infiltration showed by Coll (Fig. 5.16a) and Coll +VEGF group (Fig. 5.16c) was the lowest. It was evident from these results that the heparinized implants with VEGF demonstrated higher contraction and cellular infiltration than the non-heparinized implants.

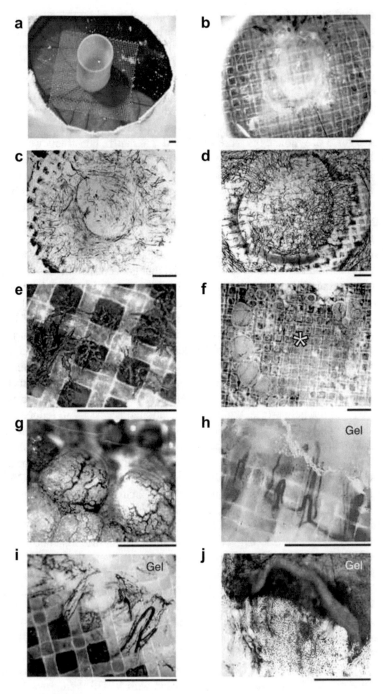

Fig. 5.14 (**a**) A fibrin and collagen gel placed on the CAM. (**b**) Response to FGF-2; the gel contracted and its opacity increased. Vascular ingrowth is visible in the periphery. (**c**) Neovessels in the

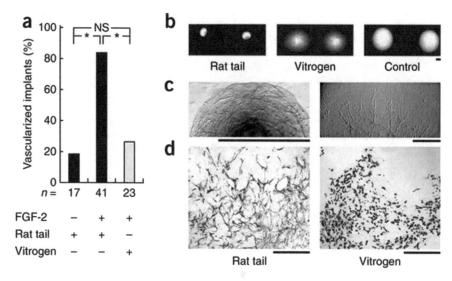

Fig. 5.15 (**a**) Inhibition of FGF-2-induced matrix neovascularization by use of Vitrogen instead of rat tail collagen ($^*P < 0.0001$). (**b, c**) In vitro cultures of myofibroblasts mixed with rat tail collagen were contracted (**b**) by cells that grew concentrically (**c**, left). Vitrogen cultures showed minimal contraction (**b**), with cells growing radially (**c**, right). Control, initial size of the droplet containing cells and collagen. Representative of $n = 35–38$ gels for each condition. (**d**) Staining for αSMA. Collagen plus fibrin gels placed on the CAM were populated by αSMA+ myofibroblasts that formed branched structures in rat tail collagen plus fibrin gels (left; representative of $n = 12$ experiments) but appeared as rounded cells in Vitrogen plus fibrin gels (right; representative of $n = 8$ experiments). (Figure adapted from Kilarski et al. [123] with permission)

The cellularization of VEGF-loaded collagen constructs can be a result of the VEGF signaling gradient present across the boundary of the collagen implant and CAM. Physiologically, such a VEGF gradient is responsible for attracting endothelial sprouts toward hypoxic regions in diseased tissues [124], exercising skeletal

---

Fig. 5.14 (continued) ingrown tissue, stained for hemoglobin. (**d, e**) Vascularized implants after i.v. injection of India ink and staining for hemoglobin to discriminate functional vessels (black, sometimes appearing greenish) from blunt-ended or under-perfused ones (red-brown). (**f**) The ingrowing vasculature was embedded in CAM tissue and appeared first at the implant periphery. Asterisk, center of implanted gel. (**g**) Enlargement of a growing "tissue bud"; it acquired its shape while growing through the mesh. (**h, i**) Vessels elongated within the growing tissue and formed functionally perfused macrovascular loops supporting a capillary network at the front of the ingrowing tissue. Blood perfusion in the live embryo, **h**; ink injection to mark functional circulation, (**i**) Ingrowing neovessels formed a clear interface to the gel. (**j**) Staining of the gel with ink. Vessels did not enter the gel but were contained within a new stain-free matrix that replaced the implanted gel ($n = 12$). Hemoglobin stained with DAB in **c–g, i, j**. Vasculature was perfused with ink in **d, e, i**. In **c–e, g, i, j** the tissue and matrix were rendered transparent by BBBA treatment. Results in **b–e, h, i** are based on at least 20 observations. Scale bars, 1 mm. (Figure adapted from Kilarski et al. [123] with permission)

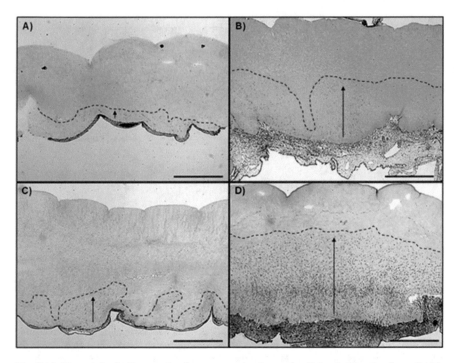

Fig. 5.16 Heparinized oligomer implants support enhanced cellular infiltration from CAM.
H&E staining of transverse section of CAM implanted with high-fibril density (20 mg/ml) implants
for 3 days (ED 10, 11, and 12) consisting of (**A**) **oligomer collagen only,** (**B**) **oligomer + 1 μg/ml
heparin,** (**C**) **oligomer + 0.5 μg/ml VEGF,** (**D**) **oligomer + 1 μg/ml heparin + 0.5 μg/ml VEGF.**
Histology reveals cellular infiltration from CAM into the collagen implants. The dotted line indicates the approximate distance in collagen up to which the CAM Cells infiltrated the collagen.
Arrows indicate the direction of cellular infiltration. Scale bar represents 50 μM. (Figure is based
on own unpublished work)

muscle [124, 125], and wounds [126]. A putative gradient of VEGF formed in collagen implants has also been reported previously to be responsible for cell recruitment in other studies involving VEGF [67] and other growth factor delivery through
collagen implants [127].

The period for which CAM is implanted with collagen also affects the cellularization and contraction of the collagen implant. For example, by leaving the collagen implants without VEGF for 9 days postimplantation instead of 3 days, a
heightened cellular infiltration and contraction of collagen can be observed
(Fig. 5.17). Oligomer collagen implants on EDD 9 are seen in Fig. 5.17a, c. The
collagen contraction post 9 days (Fig. 5.17d) is significantly increased compared to
contraction observed post 3 days of implantation (Fig. 5.17b). When scored from
the top view, the CAM shows more vascularity after 9 days of implantation compared
to 3 days of implantation (Fig. 5.17e). Histology-based H&E staining of the implants
indicates that the cell infiltration is enhanced post-9-day implantation (Fig. 5.17h)
compared to the 3-day implantation (Fig. 5.17g). Moreover, the morphology of

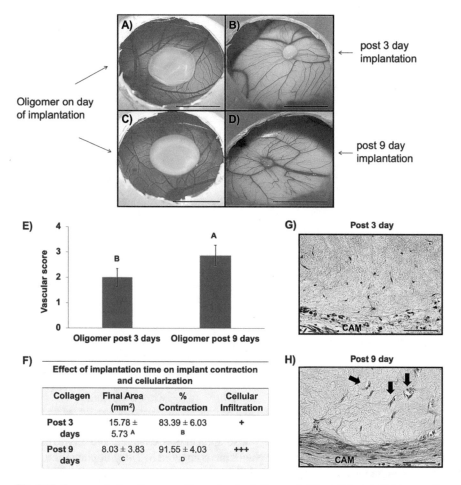

Fig. 5.17 Increase in the oligomer collagen implantation period from 3 days to 9 days results in an enhanced vascular response of CAM along with increased cell infiltration, and neovascularization within the implants. 3 mg/ml oligomer implants on the day of insertion (**A** and **C**), post-3-day implantation (**B**), and post 9-day implantation (**D**). Percent contraction of implant area (mm²) was quantified based on the differences between the original and final area of implants (**F**). H&E staining of transverse histological sections of CAM revealed enhanced cell infiltration and capillary formation post-9-day implantation (**H**) compared to post 3-day implantation (**G**). Black arrows indicate capillary formation (**H**). Scale bar represents 10 mm in a-d, and 50 µM in (**G** and **H**). Letters in (**E** and **F**) indicate statistically different experimental groups as determined by the Tukey-Kramer range test ($N = 6$–8, $p < 0.05$)

CAM cells that move into the collagen indicate improvement of cell health and size, and the cells appear to be actively involved in creating capillary sized lumens inside collagen. The implant cross-sectional area can be measured to study the contraction of collagen. The contraction of collagen by area measurement seems significantly more post 9 days of implantation compared to 3 days of implantation (Fig. 5.17f).

Fig. 5.18 Schematic representation of the collagen matrix in the mouse model and CAM model. (a) Collagen implant placed in a mouse model is covered on three sides by mouse tissue, thus holding the graft in place while cells infiltrate. In contrast, implants are not held in place in the CAM model, and as proto- and myofibroblasts migrate into the collagen matrix, the implant can contract

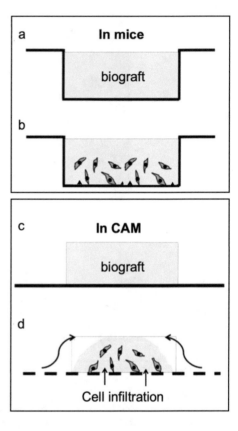

Observation of the CAM histology also indicates a changed and more mature appearance of CAM post-9-day implantation compared to 3-day implantation. This observation emphasizes the need to deliver growth factors to accelerate the vascularization of collagen implants, a lack of which takes much longer time frames to vascularize the collagen implants.

It is noteworthy mentioning here that the contraction observed in collagen implants described herein is a result of placing the inserts on CAM. If the implants were placed in a different animal model such as mice, the contraction might be minimal due to the implant being held in place from various sides by the native tissue, as represented in Fig. 5.18.

I would like to point out here that despite several successful studies performed by scientists on collagen implants in CAM and rat subcutaneous models, one lacuna remains. The period required for cellularization of the collagen constructs is still large (greater than at least 3 days) [1, 52–57]. In light of the current state-of-the-art multifunctional collagen platform developments, if there is potential to induce vasculature in collagen in a shorter time frame, these new collagen construct formulations need to be validated for evidence of distinct microvasculature as well as cellularization inside them using CAM or animal models.

5.10   Where to, After the CAM Model?

The outcome of the CAM model study would be immensely valuable in determining the safety and efficacy of the collagen biografts and would pave the way forward for future animal studies. It helps in the translation of collagen products to address many unmet clinical needs. The clinical translation of collagen-based soft tissue biografts will depend on an interactive, back and forth, "bedside to bench and back again" approach that has recently emerged. Many indispensable steps should be met within this approach, the first and foremost being *in vivo* trials using small and large animal models to evaluate the safety and efficacy of the biografts for the desired clinical need. For the designer collagen biografts to reach patient care soon, such an approach is of utmost importance.

Among the various models that can be used for this purpose, such as rabbits, dogs, goats, sheep, or pigs, the pig model is good for studies for future work, because of its known anatomical and physiological similarities to humans [128]. The animal model experiments need to be carefully designed to include variables in the system and a study should be designed to test each variable separately. As required by law, the animal research protocol would then need to be submitted to the Institutional Animal Care and Use Committee (IACUC) for approval. Any work involving animal trials should satisfy the stringent ethical considerations first. These ethical conditions include, for example, that the tissue-engineered product must have undergone sufficient testing *in vitro* and in animals such that it is likely to be safe in humans, and the tissue engineering approach should have generated sufficient data in the clinical trials before introducing into population-wide clinical practice. Depending on the success of animal trials, further stringent steps should be met, including but not limited to the human clinical trials, filing for regulatory approval by the US Food and Drug Administration (FDA), commercialization, and acceptance through physician culture change.

5.11   Summary

Physiologically, the process of vessel formation takes place in the ECM, which constitutes a dynamic 3D microenvironment of cells, providing the instructive biomechanical and biomolecular signaling required for morphogenesis. The ECM is the natural biological material, which with the help of molecules such as heparin sulfate proteoglycans or heparin, regulates the sprouting of new blood vessels, and their stabilization, leading to the restoration of functional blood circulation into ischemic tissues. Inspired by this role of ECM and heparin in spatiotemporal regulation of growth factors in vivo, new physiologically relevant collagen implants can be designed that can control the local presentation and release of VEGF at the site of implantation. Heparin's affinity for collagen molecules and VEGF189 can be leveraged for this purpose, enabling longer retention of VEGF189 in the collagen implants for the promotion of vascularization.

While CAM assay allows us to evaluate the viability of collagen implants as an angiogenic biomaterial in a rapid, simple, and low-cost *in vivo* setting, it should be noted that this model system is an intermediate step between cell culture and large animal studies or more a complex mammalian model. Therefore, the positive results of enhanced vascularization through heparinization of collagen implants obtained in any study must be tested in a large animal and mammalian model, and the differences between avian and mammalian biology should be taken into account before applying any conclusions from CAM assay to a mammalian model [129].

Overall, it is important to design collagen implants that retain collagen's multiscale structural features and inherent biological signaling capacity while promoting microvasculature formation inside the implants in an accelerated manner. Accelerated vascularization in turn can shorten the time of cellularization of constructs, decrease the risk of infection, and result in faster tissue integration and regeneration or healing of the affected tissue [130], thus providing an ideal platform for integrated tissue engineering and molecular therapy design.

References

1. C. Yao, M. Markowicz, N. Pallua, E. Magnus Noah, G. Steffens, The effect of cross-linking of collagen matrices on their angiogenic capability. Biomaterials **29**, 66–74 (2008)
2. Z. Li, T. Qu, C. Ding, C. Ma, H. Sun, S. Li, X. Liu, Injectable gelatin derivative hydrogels with sustained vascular endothelial growth factor release for induced angiogenesis. Acta Biomater. **13**, 88–100 (2015)
3. J.S. Pieper, T. Hafmans, P.B. van Wachem, M.J.A. van Luyn, L.A. Brouwer, J.H. Veerkamp, T.H. van Kuppevelt, Loading of collagen-heparan sulfate matrices with bFGF promotes angiogenesis and tissue generation in rats. J. Biomed. Mater. Res. **62**, 185–194 (2002)
4. D.S. Thoma, N. Nänni, G.I. Benic, F.E. Weber, C.H.F. Hämmerle, R.E. Jung, Effect of platelet-derived growth factor-BB on tissue integration of cross-linked and non-cross-linked collagen matrices in a rat ectopic model. Clin. Oral Implants Res. **26**, 263–270 (2015)
5. K. Lee, E.A. Silva, D.J. Mooney, Growth factor delivery-based tissue engineering: General approaches and a review of recent developments. J. R. Soc. Interface **8**, 153–170 (2011)
6. S. Barrientos, H. Brem, O. Stojadinovic, M. Tomic-Canic, Clinical application of growth factors and cytokines in wound healing. Wound Repair Regen. **22**, 569–578 (2014)
7. K.E. Johnson, T.A. Wilgus, Vascular endothelial growth factor and angiogenesis in the regulation of cutaneous wound repair. Adv. Wound Care **3**, 647–661 (2014)
8. N. Ferrara, Vascular endothelial growth factor. Eur. J. Cancer **32**, 2413–2422 (1996)
9. G. Lauer, S. Sollberg, M. Cole, I. Flamme, J. Sturzebecher, K. Mann, T. Krieg, S.A. Eming, Expression and proteolysis of vascular endothelial growth factor is increased in chronic wounds. J. Invest. Dermatol. **115**, 12–18 (2000)
10. W.J. Jeffcoate, K.G. Harding, Diabetic foot ulcers. Lancet **361**, 1545–1551 (2003)
11. Z. Wang, Z. Wang, W.W. Lu, W. Zhen, D. Yang, S. Peng, Novel biomaterial strategies for controlled growth factor delivery for biomedical applications. NPG Asia Mater **9**, e435–e435 (2017)
12. S. Guo, L.A. DiPietro, Factors affecting wound healing. J. Dent. Res. **89**, 219–229 (2010)

13. F.-M. Chen, Y. An, R. Zhang, M. Zhang, New insights into and novel applications of release technology for periodontal reconstructive therapies. J. Control. Release **149**, 92–110 (2011)
14. R.R. Chen, D.J. Mooney, Polymeric growth factor delivery strategies for tissue engineering. Pharm. Res. **20**, 1103–1112 (2003)
15. P.S. Briquez, J.A. Hubbell, M.M. Martino, Extracellular matrix-inspired growth factor delivery systems for skin wound healing. Adv. Wound Care **4**, 479–489 (2015)
16. M.M. Martino, S. Brkic, E. Bovo, M. Burger, D.J. Schaefer, T. Wolff, L. Gürke, P.S. Briquez, H.M. Larsson, R. Gianni-Barrera, J.A. Hubbell, A. Banfi, Extracellular matrix and growth factor engineering for controlled angiogenesis in regenerative medicine. Front. Bioeng. Biotechnol. **3**, 45 (2015)
17. D.L. Rabenstein, Heparin and heparan sulfate: Structure and function. Nat. Prod. Rep. **19**, 312–331 (2002)
18. S. Tessler, P. Rockwell, D. Hicklin, T. Cohen, B.Z. Levi, L. Witte, I.R. Lemischka, G. Neufeld, Heparin modulates the interaction of VEGF165 with soluble and cell associated flk-1 receptors. J. Biol. Chem. **269**, 12456–12461 (1994)
19. H. Gitay-Goren, S. Soker, I. Vlodavsky, G. Neufeld, The binding of vascular endothelial growth factor to its receptors is dependent on cell surface-associated heparin-like molecules. J. Biol. Chem. **267**, 6093–6098 (1992)
20. C. Ruhrberg, H. Gerhardt, M. Golding, R. Watson, S. Ioannidou, H. Fujisawa, C. Betsholtz, D.T. Shima, Spatially restricted patterning cues provided by heparin-binding VEGF-A control blood vessel branching morphogenesis. Genes Dev. **16**, 2684–2698 (2002)
21. H. Gerhardt, M. Golding, M. Fruttiger, C. Ruhrberg, A. Lundkvist, A. Abramsson, M. Jeltsch, C. Mitchell, K. Alitalo, D. Shima, C. Betsholtz, VEGF guides angiogenic sprouting utilizing endothelial tip cell filopodia. J. Cell Biol. **161**, 1163–1177 (2003)
22. N. Ferrara, Binding to the extracellular matrix and proteolytic processing: Two key mechanisms regulating vascular endothelial growth factor action. Mol. Biol. Cell **21**, 687–690 (2010)
23. P. Vempati, A.S. Popel, F. Mac Gabhann, Formation of VEGF isoform-specific spatial distributions governing angiogenesis: Computational analysis. BMC Syst. Biol. **5**, 59 (2011)
24. P.S. Briquez, L.E. Clegg, M.M. Martino, F.M. Gabhann, J.A. Hubbell, Design principles for therapeutic angiogenic materials. Nat. Rev. Mater. **1**, 15006 (2016)
25. P.M. Kharkar, K.L. Kiick, A.M. Kloxin, Designing degradable hydrogels for orthogonal control of cell microenvironments. Chem. Soc. Rev. **42**, 7335–7372 (2013)
26. S.E. Sakiyama-Elbert, Incorporation of heparin into biomaterials. Acta Biomater. **10**, 1581–1587 (2014)
27. K. Vulic, M.S. Shoichet, Affinity-based drug delivery systems for tissue repair and regeneration. Biomacromolecules **15**, 3867–3880 (2014)
28. Y. Liang, K.L. Kiick, Heparin-functionalized polymeric biomaterials in tissue engineering and drug delivery applications. Acta Biomater. **10**, 1588–1600 (2014)
29. Y.K. Joung, J.W. Bae, K.D. Park, Controlled release of heparin-binding growth factors using heparin-containing particulate systems for tissue regeneration. Expert Opin. Drug Deliv. **5**, 1173–1184 (2008)
30. N. Ferrara, K.A. Houck, L.B. Jakeman, J. Winer, D.W. Leung, The vascular endothelial growth factor family of polypeptides. J. Cell. Biochem. **47**, 211–218 (1991)
31. S.M. Sweeney, C.A. Guy, G.B. Fields, J.D. San Antonio, Defining the domains of type I collagen involved in heparin- binding and endothelial tube formation. Proc. Natl. Acad. Sci. U. S. A. **95**, 7275–7280 (1998)
32. J.D. San Antonio, A.D. Lander, M.J. Karnovsky, H.S. Slayter, Mapping the heparin binding sites on type I collagen monomers and fibrils. J. Cell Biol. **125**, 1179–1188 (1994)
33. K.M. Keller, J.M. Keller, K. Kühn, The C-terminus type I collagen is a major binding site for heparin. Biochim. Biophys. Acta Gen. Subj. **882**, 1–5 (1986)

34. D.R. Stamov, T.A. Khoa Nguyen, H.M. Evans, T. Pfohl, C. Werner, T. Pompe, The impact of heparin intercalation at specific binding sites in telopeptide-free collagen type I fibrils. Biomaterials **32**, 7444–7453 (2011)
35. M.B. Mathews, L. Decker, The effect of acid mucopolysaccharides and acid mucopolysaccharide-proteins on fibril formation from collagen solutions. Biochem. J. **109**, 517–526 (1968)
36. G.C. Wood, The formation of fibrils from collagen solutions. 3. Effect of chondroitin sulphate and some other naturally occurring polyanions on the rate of formation. Biochem. J. **75**, 605–612 (1960)
37. C. Guidry, F. Grinnell, Heparin modulates the organization of hydrated collagen gels and inhibits gel contraction by fibroblasts. J. Cell Biol. **104**, 1097–1103 (1987)
38. J.M. McPherson, S.J. Sawamura, R.A. Condell, W. Rhee, D.G. Wallace, The effects of heparin on the physicochemical properties of reconstituted collagen. Coll. Relat. Res. **8**, 65–82 (1988)
39. B. Mulloy, M. Forster, C. Jones, D. Davies, Nmr and molecular-modelling studies of the solution conformation of heparin. Biochem. J. **293**, 849–858 (1993)
40. D. Krilleke, Y.S. Ng, D.T. Shima, The heparin-binding domain confers diverse functions of VEGF-A in development and disease: A structure-function study. Biochem. Soc. Trans. **37**, 1201–1206 (2009)
41. M. Tillo, L. Erskine, A. Cariboni, A. Fantin, A. Joyce, L. Denti, C. Ruhrberg, VEGF189 binds NRP1 and is sufficient for VEGF/NRP1-dependent neuronal patterning in the developing brain. Development **142**, 314–319 (2015)
42. F.M. Gabhann, A.S. Popel, Systems biology of vascular endothelial growth factors. Microcirculation **15**, 715–738 (2008)
43. G. Neufeld, T. Cohen, H. Gitay-Goren, Z. Poltorak, S. Tessler, R. Sharon, S. Gengrinovitch, B.-Z. Levi, Similarities and differences between the vascular endothelial growth factor (VEGF) splice variants. Cancer Metastasis Rev. **15**, 153–158 (1996)
44. D.W. Leung, G. Cachianes, W.J. Kuang, D.V. Goeddel, N. Ferrara, Vascular endothelial growth factor is a secreted angiogenic mitogen. Science **246**, 1306 (1989)
45. J.E. Park, G.A. Keller, N. Ferrara, The vascular endothelial growth factor (VEGF) isoforms: Differential deposition into the subepithelial extracellular matrix and bioactivity of extracellular matrix-bound VEGF. Mol. Biol. Cell **4**, 1317–1326 (1993)
46. T. Cohen, H. Gitay-Goren, R. Sharon, M. Shibuya, R. Halaban, B. Levi, G. Neufeld, VEGF121, a vascular endothelial growth factor (VEGF) isoform lacking heparin binding ability, requires cell-surface heparan sulfates for efficient binding to the VEGF receptors of human melanoma cells. J. Biol. Chem. **270**, 11322–11326 (1995)
47. N. Vaisman, D. Gospodarowicz, G. Neufeld, Characterization of the receptors for vascular endothelial growth factor. J. Biol. Chem. **265**, 19461–19466 (1990)
48. M. Shibuya, Vascular endothelial growth factor (VEGF) and its receptor (VEGFR) signaling in angiogenesis: A crucial target for anti- and pro-angiogenic therapies. Genes Cancer **2**, 1097–1105 (2011)
49. S. Ashikari-Hada, H. Habuchi, Y. Kariya, K. Kimata, Heparin regulates vascular endothelial growth factor 165-dependent Mitogenic activity, tube formation, and its receptor phosphorylation of human endothelial cells: Comparison of the effects of heparin and modified heparins. J. Biol. Chem. **280**, 31508–31515 (2005)
50. K.A. Houck, D.W. Leung, A.M. Rowland, J. Winer, N. Ferrara, Dual regulation of vascular endothelial growth factor bioavailability by genetic and proteolytic mechanisms. J. Biol. Chem. **267**, 26031–26037 (1992)
51. D. Krilleke, Y.-S.E. Ng, D.T. Shima, The heparin-binding domain confers diverse functions of VEGF-A in development and disease: A structure–function study. Biochem. Soc. Trans. **37**, 1201 (2009)

52. J.-H. Kim, T.-H. Kim, M.S. Kang, H.-W. Kim, Angiogenic effects of collagen/mesoporous nanoparticle composite scaffold delivering VEGF(165). Biomed. Res. Int. **2016**, 9676934 (2016)
53. S.T.M. Nillesen, P.J. Geutjes, R. Wismans, J. Schalkwijk, W.F. Daamen, T.H. van Kuppevelt, Increased angiogenesis and blood vessel maturation in acellular collagen–heparin scaffolds containing both FGF2 and VEGF. Biomaterials **28**, 1123–1131 (2007)
54. G.C. Steffens, C. Yao, P. Prevel, M. Markowicz, P. Schenck, E.M. Noah, N. Pallua, Modulation of angiogenic potential of collagen matrices by covalent incorporation of heparin and loading with vascular endothelial growth factor. Tissue Eng. **10**, 1502–1509 (2004)
55. C. Yao, P. Prével, S. Koch, P. Schenck, E.M. Noah, N. Pallua, G. Steffens, Modification of collagen matrices for enhancing angiogenesis. Cells Tissues Organs **178**, 189–196 (2004)
56. G. Grieb, A. Groger, A. Piatkowski, M. Markowicz, G.C.M. Steffens, N. Pallua, Tissue substitutes with improved angiogenic capabilities: An in vitro investigation with endothelial cells and endothelial progenitor cells. Cells Tissues Organs **191**, 96–104 (2010)
57. K.M. Brouwer, R.M. Wijnen, D. Reijnen, T.G. Hafmans, W.F. Daamen, T.H. van Kuppevelt, Heparinized collagen scaffolds with and without growth factors for the repair of diaphragmatic hernia. Organogenesis **9**, 161–167 (2013)
58. C. Yao, M. Roderfeld, T. Rath, E. Roeb, J. Bernhagen, G. Steffens, The impact of proteinase-induced matrix degradation on the release of VEGF from heparinized collagen matrices. Biomaterials **27**, 1608–1616 (2006)
59. K.S. Weadock, E.J. Miller, L.D. Bellincampi, J.P. Zawadsky, M.G. Dunn, Physical crosslinking of collagen fibers: Comparison of ultraviolet irradiation and dehydrothermal treatment. J. Biomed. Mater. Res. **29**, 1373–1379 (1995)
60. L. Ma, C. Gao, Z. Mao, J. Zhou, J. Shen, Enhanced biological stability of collagen porous scaffolds by using amino acids as novel cross-linking bridges. Biomaterials **25**, 2997–3004 (2004)
61. J.-H. Kim, T.-H. Kim, M.S. Kang, H.-W. Kim, Angiogenic effects of collagen/mesoporous nanoparticle composite scaffold delivering VEGF165. Biomed. Res. Int. **2016**, 8 (2016)
62. S.M. Jay, B.R. Shepherd, J.W. Andrejecsk, T.R. Kyriakides, J.S. Pober, W.M. Saltzman, Dual delivery of VEGF and MCP-1 to support endothelial cell transplantation for therapeutic vascularization. Biomaterials **31**, 3054–3062 (2010)
63. E. Quinlan, A. López-Noriega, E.M. Thompson, A. Hibbitts, S.A. Cryan, F.J. O'Brien, Controlled release of vascular endothelial growth factor from spray-dried alginate microparticles in collagen–hydroxyapatite scaffolds for promoting vascularization and bone repair. J. Tissue Eng. Regen. Med. (2015) n/a-n/a
64. N. Nagai, N. Kumasaka, T. Kawashima, H. Kaji, M. Nishizawa, T. Abe, Preparation and characterization of collagen microspheres for sustained release of VEGF. J. Mater. Sci. Mater. Med. **21**, 1891–1898 (2010)
65. I. Wilcke, J.A. Lohmeyer, S. Liu, A. Condurache, S. Kruger, P. Mailander, H.G. Machens, VEGF(165) and bFGF protein-based therapy in a slow release system to improve angiogenesis in a bioartificial dermal substitute in vitro and in vivo. Langenbeck's Arch. Surg. **392**, 305–314 (2007)
66. Y.H. Shen, M.S. Shoichet, M. Radisic, Vascular endothelial growth factor immobilized in collagen scaffold promotes penetration and proliferation of endothelial cells. Acta Biomater. **4**, 477–489 (2008)
67. D. Odedra, L.L.Y. Chiu, M. Shoichet, M. Radisic, Endothelial cells guided by immobilized gradients of vascular endothelial growth factor on porous collagen scaffolds. Acta Biomater. **7**, 3027–3035 (2011)
68. L.L.Y. Chiu, M. Radisic, Scaffolds with covalently immobilized VEGF and Angiopoietin-1 for vascularization of engineered tissues. Biomaterials **31**, 226–241 (2010)
69. Y. Tabata, M. Miyao, M. Ozeki, Y. Ikada, Controlled release of vascular endothelial growth factor by use of collagen hydrogels. J. Biomater. Sci. Polym. Ed. **11**, 915–930 (2000)

70. Q. He, Y. Zhao, B. Chen, Z. Xiao, J. Zhang, L. Chen, W. Chen, F. Deng, J. Dai, Improved cellularization and angiogenesis using collagen scaffolds chemically conjugated with vascular endothelial growth factor. Acta Biomater. **7**, 1084–1093 (2011)
71. Y. Miyagi, L.L.Y. Chiu, M. Cimini, R.D. Weisel, M. Radisic, R.-K. Li, Biodegradable collagen patch with covalently immobilized VEGF for myocardial repair. Biomaterials **32**, 1280–1290 (2011)
72. K. Raghunath, G. Biswas, K.P. Rao, K.T. Joseph, M. Chvapil, Some characteristics of collagen-heparin complex. J. Biomed. Mater. Res. **17**, 613–621 (1983)
73. M.B. Mathews, L. Decker, The effect of acid mucopolysaccharides and acid mucopolysaccharide–proteins on fibril formation from collagen solutions. Biochem. J. **109**, 517–526 (1968)
74. D. Stamov, M. Grimmer, K. Salchert, T. Pompe, C. Werner, Heparin intercalation into reconstituted collagen I fibrils: Impact on growth kinetics and morphology. Biomaterials **29**, 1–14 (2008)
75. A. Lode, A. Reinstorf, A. Bernhardt, C. Wolf-Brandstetter, U. König, M. Gelinsky, Heparin modification of calcium phosphate bone cements for VEGF functionalization. J. Biomed. Mater. Res. A **86A**, 749–759 (2008)
76. C. Wolf-Brandstetter, A. Lode, T. Hanke, D. Scharnweber, H. Worch, Influence of modified extracellular matrices on TI6AL4V implants on binding and release of VEGF. J. Biomed. Mater. Res. A **79A**, 882–894 (2006)
77. S. Knaack, A. Lode, B. Hoyer, A. Rösen-Wolff, A. Gabrielyan, I. Roeder, M. Gelinsky, Heparin modification of a biomimetic bone matrix for controlled release of VEGF. J. Biomed. Mater. Res. A **102**, 3500–3511 (2014)
78. U. Konig, A. Lode, P.B. Welzel, Y. Ueda, S. Knaack, A. Henss, A. Hauswald, M. Gelinsky, Heparinization of a biomimetic bone matrix: Integration of heparin during matrix synthesis versus adsorptive post surface modification. J. Mater. Sci. Mater. Med. **25**, 607–621 (2014)
79. Y.-T. Hou, H. Ijima, T. Takei, K. Kawakami, Growth factor/heparin-immobilized collagen gel system enhances viability of transplanted hepatocytes and induces angiogenesis. J. Biosci. Bioeng. **112**, 265–272 (2011)
80. R. Joshi, Designer collagen-fibril biograft materials for tunable molecular delivery, Open Access Dissertations, in: Biomedical Engineering, Purddue University, West Lafayette, Inddiana, 2016
81. A. Vargas, M. Zeisser-Labouèbe, N. Lange, R. Gurny, F. Delie, The chick embryo and its chorioallantoic membrane (CAM) for the in vivo evaluation of drug delivery systems. Adv. Drug Deliv. Rev. **59**, 1162–1176 (2007)
82. D. Ribatti, B. Nico, A. Vacca, L. Roncali, P.H. Burri, V. Djonov, Chorioallantoic membrane capillary bed: A useful target for studying angiogenesis and anti-angiogenesis in vivo. Anat. Rec. **264**, 317–324 (2001)
83. T. Leng, J.M. Miller, K.V. Bilbao, D.V. Palanker, P. Huie, M.S. Blumenkranz, The chick chorioallantoic membrane as a model tissue for surgical retinal research and simulation. Retina **24**, 427–434 (2004)
84. D. Ribatti, *The Chick Embryo Chorioallantoic Membrane in the Study of Angiogenesis and Metastasis: The CAM Assay in the Study of Angiogenesis and Metastasis* (Springer, Dordrecht, 2010)
85. P. Nowak-Sliwinska, T. Segura, M.L. Iruela-Arispe, The chicken chorioallantoic membrane model in biology, medicine and bioengineering. Angiogenesis **17**, 779–804 (2014)
86. D.H. Ausprunk, D.R. Knighton, J. Folkman, Vascularization of normal and neoplastic tissues grafted to the chick chorioallantois. Role of host and preexisting graft blood vessels. Am. J. Pathol. **79**, 597–628 (1975)
87. M.L. Ponce, H.K. Kleinmann, The chick chorioallantoic membrane as an in vivo angiogenesis model, in *Current Protocols in Cell Biology*, (Wiley, New York, 2001)
88. ICCVAM-recommended test method protocol: Hen's egg test – Chorioallantoic membrane (HET-CAM) test method, in: NIH Publication No. 10–7553, NIH, 2010

89. D.C. West, W.D. Thompson, P.G. Sells, M.F. Burbridge, Angiogenesis assays using chick chorioallantoic membrane. Methods Mol. Med. **46**, 107–129 (2001)
90. X. Liu, X. Wang, A. Horii, L. Qiao, S. Zhang, F.Z. Cui, In vivo studies on angiogenic activity of two designer self-assembling peptide scaffold hydrogels in the chicken embryo chorioallantoic membrane. Nanoscale **4**, 2720–2727 (2012)
91. D.S. Dohle, S.D. Pasa, S. Gustmann, M. Laub, J.H. Wissler, H.P. Jennissen, N. Dünker, Chick ex ovo culture and ex ovo CAM assay: How it really works. J. Vis. Exp. **33**, e1620 (2009)
92. D.R. Harland, L.D. Lorenz, K. Fay, B.E. Dunn, S.K. Gruenloh, J. Narayanan, E.R. Jacobs, M. Medhora, Acute effects of prostaglandin E1 and E2 on vascular reactivity and blood flow in situ in the chick chorioallantoic membrane. Prostaglandins Leukot. Essent. Fat. Acids **87**, 79–89 (2012)
93. C. Cid Maria, J. Hernández-Rodríguez, M.-J. Esteban, M. Cebrián, S. Gho Yong, C. Font, A. Urbano-Márquez, M. Grau Josep, K. Kleinman Hynda, Tissue and serum angiogenic activity is associated with low prevalence of ischemic complications in patients with giant-cell arteritis. Circulation **106**, 1664–1671 (2002)
94. D. Ribatti, B. Nico, A. Vacca, M. Presta, The gelatin sponge-chorioallantoic membrane assay. Nat. Protoc. **1**, 85–91 (2006)
95. D. Ribatti, Chicken chorioallantoic membrane angiogenesis model, in *Cardiovascular Development: Methods and Protocols*, ed. by X. Peng, M. Antonyak, (Humana Press, Totowa, 2012), pp. 47–57
96. D. Ribatti, The chick embryo chorioallantoic membrane (CAM). A multifaceted experimental model. Mech. Dev. **141**, 70–77 (2016)
97. D. Ribatti, B. Nico, A. Vacca, L. Roncali, P.H. Burri, V. Djonov, Chorioallantoic membrane capillary bed: A useful target for studying angiogenesis and anti-angiogenesis in vivo. Anat. Rec. **264**, 317–324 (2001)
98. R.B. Rema, K. Rajendran, M. Ragunathan, Angiogenic efficacy of heparin on chick chorioallantoic membrane. Vascular Cell **4**, 8–8 (2012)
99. D. Ribatti, L. Roncali, B. Nico, M. Bertossi, Effects of exogenous heparin on the vasculogenesis of the chorioallantoic membrane. Acta Anat. (Basel) **130**, 257–263 (1987)
100. S. Pacini, M. Gulisano, S. Vannucchi, M. Ruggiero, Poly-l-lysine/heparin stimulates angiogenesis in chick embryo chorioallantoic membrane. Biochem. Biophys. Res. Commun. **290**, 820–823 (2002)
101. Q. Tan, H. Tang, J. Hu, Y. Hu, X. Zhou, Y. Tao, Z. Wu, Controlled release of chitosan/heparin nanoparticle-delivered VEGF enhances regeneration of decellularized tissue-engineered scaffolds. Int. J. Nanomedicine **6**, 929–942 (2011)
102. M.J. Wissink, R. Beernink, J.S. Pieper, A.A. Poot, G.H. Engbers, T. Beugeling, W.G. van Aken, J. Feijen, Immobilization of heparin to EDC/NHS-crosslinked collagen. Characterization and in vitro evaluation. Biomaterials **22**, 151–163 (2001)
103. M.J.B. Wissink, R. Beernink, N.M. Scharenborg, A.A. Poot, G.H.M. Engbers, T. Beugeling, W.G. van Aken, J. Feijen, Endothelial cell seeding of (heparinized) collagen matrices: Effects of bFGF pre-loading on proliferation (after low density seeding) and pro-coagulant factors. J. Control. Release **67**, 141–155 (2000)
104. W.J. Fairbrother, M.A. Champe, H.W. Christinger, B.A. Keyt, M.A. Starovasnik, Solution structure of the heparin-binding domain of vascular endothelial growth factor. Structure **6**, 637–648 (1998)
105. M.S. Ågren, C.J. Taplin, J.F. Woessner, W.H. Eagistein, P.M. Mertz, Collagenase in wound healing: Effect of wound age and type. J. Investig. Dermatol. **99**, 709–714 (1992)
106. K.G. Harding, H.L. Morris, G.K. Patel, Healing chronic wounds. BMJ **324**, 160 (2002)
107. K.M. Blum, T. Novak, L. Watkins, C.P. Neu, J.M. Wallace, Z.R. Bart, S.L. Voytik-Harbin, Acellular and cellular high-density, collagen-fibril constructs with suprafibrillar organization. Biomater. Sci. **4**, 711–723 (2016)
108. W. Friess, Collagen – Biomaterial for drug delivery. Eur. J. Pharm. Biopharm. **45**, 113–136 (1998)

109. M. Miron-Mendoza, J. Seemann, F. Grinnell, The differential regulation of cell motile activ-
     ity through matrix stiffness and porosity in three dimensional collagen matrices. Biomaterials
     31, 6425–6435 (2010)
110. R.J. Lee, M.L. Springer, W.E. Blanco-Bose, R. Shaw, P.C. Ursell, H.M. Blau, VEGF gene
     delivery to myocardium. Circulation **102**, 898 (2000)
111. M.L. Springer, A.S. Chen, P.E. Kraft, M. Bednarski, H.M. Blau, VEGF gene delivery to
     muscle: Potential role for vasculogenesis in adults. Mol. Cell **2**, 549–558 (1998)
112. C.J. Drake, C.D. Little, Exogenous vascular endothelial growth factor induces malformed
     and hyperfused vessels during embryonic neovascularization. Proc. Natl. Acad. Sci. **92**,
     7657–7661 (1995)
113. A.H. Zisch, M.P. Lutolf, J.A. Hubbell, Biopolymeric delivery matrices for angiogenic growth
     factors. Cardiovasc. Pathol. **12**, 295–310 (2003)
114. M. Ehrbar, V.G. Djonov, C. Schnell, S.A. Tschanz, G. Martiny-Baron, U. Schenk, J. Wood,
     P.H. Burri, J.A. Hubbell, A.H. Zisch, Cell-demanded liberation of VEGF121 from fibrin
     implants induces local and controlled blood vessel growth. Circ. Res. **94**, 1124–1132 (2004)
115. N. Nagai, N. Kumasaka, T. Kawashima, H. Kaji, M. Nishizawa, T. Abe, Preparation and char-
     acterization of collagen microspheres for sustained release of VEGF. J. Mater. Sci. Mater.
     Med. **21**, 1891–1898 (2010)
116. S. Soker, S. Takashima, H.Q. Miao, G. Neufeld, M. Klagsbrun, Neuropilin-1 is expressed by
     endothelial and tumor cells as an isoform-specific receptor for vascular endothelial growth
     factor. Cell **92**, 735–745 (1998)
117. M.A. Princz, H. Sheardown, Heparin-modified dendrimer crosslinked collagen matrices for
     the delivery of heparin-binding epidermal growth factor. J. Biomed. Mater. Res. A **100A**,
     1929–1937 (2012)
118. R.L. Jackson, S.J. Busch, A.D. Cardin, Glycosaminoglycans: Molecular properties, protein
     interactions, and role in physiological processes. Physiol. Rev. **71**, 481–539 (1991)
119. I. Capila, R.J. Linhardt, Heparin-protein interactions. Angew. Chem. Int. Ed. Engl. **41**, 391–
     412 (2002)
120. B. ÖBrink, A study of the interactions between monomeric tropocollagen and glycosamino-
     glycans. Eur. J. Biochem. **33**, 387–400 (1973)
121. M.B. Mathews, The interaction of collagen and acid mucopolysaccharides. A model for con-
     nective tissue. Biochem. J. **96**, 710 (1965)
122. A.Y. Wang, C.A. Foss, S. Leong, X. Mo, M.G. Pomper, S.M. Yu, Spatio-temporal modi-
     fication of collagen scaffolds mediated by triple helical propensity. Biomacromolecules **9**,
     1755–1763 (2008)
123. W.W. Kilarski, B. Samolov, L. Petersson, A. Kvanta, P. Gerwins, Biomechanical regulation
     of blood vessel growth during tissue vascularization. Nat. Med. **15**, 657–664 (2009)
124. B. Hoier, M. Walker, M. Passos, P.J. Walker, A. Green, J. Bangsbo, C.D. Askew, Y. Hellsten,
     Angiogenic response to passive movement and active exercise in individuals with peripheral
     arterial disease. J. Appl. Physiol. **115**(2013), 1777–1787 (1985)
125. B. Hoier, N. Nordsborg, S. Andersen, L. Jensen, L. Nybo, J. Bangsbo, Y. Hellsten, Pro- and
     anti-angiogenic factors in human skeletal muscle in response to acute exercise and training.
     J. Physiol. **590**, 595–606 (2012)
126. C. Kut, F. Mac Gabhann, A.S. Popel, Where is VEGF in the body? A meta-analysis of VEGF
     distribution in cancer. Br. J. Cancer **97**, 978–985 (2007)
127. B.A. Bladergroen, B. Siebum, K.G.C. Siebers-Vermeulen, T.H. Van Kuppevelt, A.A. Poot,
     J. Feijen, C.G. Figdor, R. Torensma, In vivo recruitment of hematopoietic cells using stromal
     cell–derived factor 1 alpha–loaded heparinized three-dimensional collagen scaffolds. Tissue
     Eng. A **15**, 1591–1599 (2009)
128. J. Haier, F. Schmidt, In vivo animal models in tissue engineering, in *Fundamentals of Tissue
     Engineering and Regenerative Medicine*, ed. by U. Meyer, J. Handschel, H. P. Wiesmann,
     T. Meyer, (Springer Berlin Heidelberg, Berlin/Heidelberg, 2009), pp. 773–779

129. S. Baiguera, P. Macchiarini, D. Ribatti, Chorioallantoic membrane for in vivo investigation of tissue-engineered construct biocompatibility. J. Biomed. Mater. Res. B Appl. Biomater. **100B**, 1425–1434 (2012)
130. G.C. Hughes, S.S. Biswas, B. Yin, R.E. Coleman, T.R. DeGrado, C.K. Landolfo, J.E. Lowe, B.H. Annex, K.P. Landolfo, Therapeutic angiogenesis in chronically ischemic porcine myocardium: Comparative effects of bFGF and VEGF. Ann. Thorac. Surg. **77**, 812–818 (2004)

Index

Printed in the United States
By Bookmasters